T0257720

Genetic Engineering of DNA and Protein

Volume I

Genetic Engineering of DNA and Protein

Volume I

Edited by **Tom Lee**

New York

Published by Callisto Reference,
106 Park Avenue, Suite 200,
New York, NY 10016, USA
www.callistoreference.com

Genetic Engineering of DNA and Protein
Volume I
Edited by Tom Lee

International Standard Book Number: 978-1-63239-351-7 (Hardback)

Contents

Preface

This book provides readers with research works on worldwide discussions on the topic of latest molecular genetics. There are two approaches of every research work published in this book; first, to make the research chapters understandable to majority of readers and second, to describe the genetic tools and pathways used in research. The one fact mostly highlighted is the necessity of genetic insight in solving an issue. This book will prove to be an interesting read to those interested in genetic discoveries because of its structure, which has been made with a view point of attracting readers and familiarizing them with genetic approaches.

The information shared in this book is based on empirical researches made by veterans in this field of study. The elaborative information provided in this book will help the readers further their scope of knowledge leading to advancements in this field.

Finally, I would like to thank my fellow researchers who gave constructive feedback and my family members who supported me at every step of my research.

Editor

Molecular Genetics in Basic Research

1

Biochemical Analysis of Halophilic Dehydrogenases Altered by Site-Directed Mutagenesis

J. Esclapez, M. Camacho, C. Pire and M.J. Bonete
Departamento de Agroquímica y Bioquímica, División de Bioquímica y Biología Molecular,
Facultad de Ciencias, Universidad de Alicante, Alicante,
Spain

1. Introduction

Extremely halophilic Archaea are found in highly saline environments such as natural salt lakes, saltern pools, the Dead Sea and so on. These microorganisms require between 2.5 and 5.2 M NaCl for optimal growth. They can balance the external concentration by accumulating intracellular KCl to concentrations that can reach and exceed saturation. The biochemical machinery of these microorganisms has, therefore, been adapted in the course of evolution to be able to function at salt concentrations at which most biochemical systems will cease to function. The biochemical and biophysical properties of several halophilic enzymes have been studied in great detail; and, as a general rule, it was found that the halophilic enzymes are stabilized by multimolar concentration of salts. In most cases the salt also stimulates the catalytic activity. This stabilization of halophilic proteins in solvents containing high salt concentrations has been discussed in terms of apparent peculiarities in their composition. Since the first amino acid composition determinations, it has become clear that halophilic enzymes present a higher proportion of acidic over basic residues, an increase in small hydrophobic residues, a decrease in aliphatic residues and lower lysine content than their non-halophilic homologues (Lanyi, 1974; Eisenberg, et al., 1992; Madern et al., 2000). Since then, structural analyses have revealed two significant differences in the characteristics of the surface of the halophilic enzymes that may contribute to their stability in high salt. The first of these is that the excess of acidic residues are predominantly located on the enzyme surface leading to the formation of a hydration shell that protects the enzyme from aggregation in its highly saline environment. The second is that the surface also displays a significant reduction in exposed hydrophobic character, which arises not from a loss of surface exposed hydrophobic residues but from a reduction in surface-exposed lysine residues. Nevertheless, although the number of halophilic protein sequences has increased during the last years, the number of high resolution structures that permit the details of the protein solvent interactions to be seen is limited. The role of the reduction in the surface lysines has been largely ignored (Britton et al., 1998, 2006). Furthermore, in several studies, the authors have concluded that it is the precise structural organization of surface acidic residues that is important in halophilic adaptation. Not only is there an increase in acidic residue content, but these residues form clusters that bind networks of hydrated ions (Richard et al., 2000).

Halophilic archaea are considered a rather homogeneous group of heterotrophic microorganisms predominantly using amino acids as their source of carbon and energy. However, it has been shown that some halophilic archaea are able to use not only amino acids but different metabolites as well, as, for example, *Haloferax mediterranei,* which grows in a minimal medium containing glucose as the only source of carbon using a modified Entner-Doudoroff pathway (Rodriguez-Valera et al., 1983), or *Haloferax volcanii,* which is also able to grow in minimal medium with acetate as the sole carbon source (Kauri et al., 1990). Isocitrate lyase and malate synthase activities were detected in this organism when it was grown on a medium with acetate as the main carbon source (Serrano et al., 1998).

To understand the molecular basis of salt tolerance responsible for halophilic adaptation of proteins, to analyze the coenzyme specificity, and to study the mode of zinc-binding, we have chosen as model enzymes two halophilic dehydrogenase proteins involved in carbon catabolism. They are the glucose dehydrogenase (GlcDH) and isocitrate dehydrogenase (ICDH) from the extremely halophilic Archaea *Haloferax mediterranei* and *Haloferax volcanii,* respectively.

1.1 *Haloferax mediterranei* glucose dehydrogenase

GlcDH is the first enzyme of a non-phosphorylated Entner-Doudoroff pathway. It catalyses the reaction:

$$\text{Glucose} + \text{NAD(P)}^+ \rightarrow \text{Glucono-1,5-lactone} + \text{NAD(P)H} + \text{H}^+$$

GlcDH from *Hfx. mediterranei* has been characterized and purified using gel filtration and affinity chromatography in the presence of buffers containing a high concentration of salt or glycerol to stabilize its structure. The protein is a dimeric enzyme with a molecular weight of 39 kDa per subunit, and shows a dual cofactor specificity, although it displays a marked preference for NADP$^+$ to NAD$^+$. Biochemical studies have established that the presence of a divalent ion such as Mg^{2+} or Mn^{2+} at concentrations of 25 mM enhances enzymatic activity (Bonete et al., 1996). Inactivation by metal chelators and reactivation by certain divalent ions indicated that glucose dehydrogenase from *Hfx. mediterranei* contains tightly bound metal ions that are essential for activity. Studies on the metal content of the enzyme by ICP revealed the presence of zinc ions whose removal by addition of EDTA leads to complete loss of enzyme activity (Pire et al., 2000). Sequence analysis showed that this enzyme belongs to the zinc-dependent medium-chain alcohol dehydrogenase superfamily (MDR), which includes sorbitol dehydrogenases, xylitol dehydrogenases and alcohol dehydrogenases (Pire et al., 2001). The structure of *Hfx. mediterranei* GlcDH has been solved at the highest resolution to date for any water-soluble halophilic enzyme. The structures of the apoenzyme and a D38C mutant in complex with NADP$^+$ and zinc reveal that the subunit, like that of the other MDR family members, is organized into two domains separated by a deep cleft, with the active site lying at its base. Domain 1 contains the residues involved in substrate binding, catalysis, and coordination of the active-site zinc. Domain 2 consists of a dinucleotide-binding Rossmann fold (Rossmann et al., 1974) that is responsible for binding NADP$^+$. Its molecular surface is predominantly covered by acidic residues, which are only partially neutralized by bound potassium counterions that also appear to play a role in substrate binding. The surface shows the expected reduction in hydrophobic character associated with the loss of lysines, which is consistent with the genome-wide reduction of this residue in extreme halophiles. The structure also reveals a highly ordered, multilayered solvation shell that can be seen to be organized into one dominant network covering much of the exposed surface accessible area to an extent not seen in almost any other protein structure solved (Ferrer et al., 2001; Britton et al., 2006). Recently, high-resolution structures of a series of binary and ternary complexes of halophilic GlcDH have allowed an extension of the understanding of the catalytic mechanism in the MDR family. In contrast to the textbook MDR mechanism in which the zinc ion is proposed to remain stationary and attached to a common set of protein ligands, analysis of these structures reveals that in each complex, there are dramatic differences in the nature of the zinc ligation. These changes arise as a direct consequence of linked movements of the zinc ion, a zinc-bound bound water molecule, and the substrate during progression through the reaction. These results provide evidence for the molecular basis of proton traffic during catalysis, a structural explanation for pentacoordinate zinc ion intermediates, and a unifying view for the observed patterns of metal ligation in the MDR family (Esclapez et al., 2005; Baker et al., 2009).

1.2 *Haloferax volcanii* isocitrate dehydrogenase

The citric acid cycle enzyme, ICDH (EC 1.1.1.41 and EC 1.1.1.42), catalyses the oxidative decarboxylation of isocitrate (Kay & Weitzman, 1987):

$$\text{Isocitrate} + NAD(P)^+ \rightarrow \text{2-oxoglutarate} + NAD(P)H + H^+ + CO_2$$

The wild-type enzyme from *Haloferax volcanii* was purified using three steps. The enzyme has been characterized, and it is a dimer with subunit M_r of 62000 Da. Its activity is strictly NADP dependent, and markedly dependent on the concentration of NaCl or KCl, being maximal in 0.5 M NaCl or KCl. The thermostability of the archaeal isocitrate dehydrogenase was investigated incubating the enzyme in buffer containing either 0.5 M or 3 M KCl. Clearly, the thermal stability of the enzyme is substantially reduced at the lower KC1 concentration, with concomitant differences in the activation energies for the thermal inactivation process, 360 kJ mol^{-1} and 610 kJ mol^{-1} at 0.5 M and 3 M KCl, respectively; therefore, the high *in vivo* KC1 concentrations appear to be more important for the stability of the enzyme than for its catalytic ability (Camacho et al., 1995). The gene encoding this protein was sequenced and the derived amino acids were determined. The yields of *Escherichia coli*-expressed enzyme were greater than those obtained by purification of the enzyme from the native organism, but the product was insoluble inclusion bodies. The recombinant ICDH behaves similarly to the native enzyme with respect to the dependence of activity on salt concentration. Kinetic analysis has also shown the purified recombinant and native enzymes to be similar, as are the thermal stabilities (Camacho et al., 2002). *Hfx. volcanii* ICDH dissociation/deactivation has been measured to probe the respective effect of anions and cations on stability. Surprisingly, enzyme stability has been found to be mainly sensitive to cations and very little (or not) to anions. Divalent cations have induced a strong shift of the active/inactive transition towards low salt concentration. A high resistance of ICDH from *Hfx. volcanii* to chemical denaturation has also been found. This study strongly suggests that *Hfx. volcanii* ICDH might be seen as a type of halophilic protein never described before: an oligomeric halophilic protein devoid of intersubunit anion-binding sites (Madern et al., 2004).

2. Materials and methods

2.1 Strains, culture conditions and vectors

Escherichia coli NovaBlue (Novagen) was used as host for plasmids pGEM-11Zf(+) and pET3a. *E. coli* BMH71-18 *mutS* (Promega) and *E. coli* XL1-Blue (Stratagene) were employed in site-directed mutagenesis experiments. *E. coli* BL21(DE3) (Novagen) was used as the expression host. *E. coli* strains were grown in Luria-Bertani medium at 37 °C with shaking at 180 rpm. Plasmids were selected for in solid and liquid media by the addition of 100 μg ampicillin/ml.

Vector pGEM-11Zf(+) (Promega) was used for cloning genes and carrying out some site-directed mutagenesis experiments. The expression vector pET3a was purchased from Novagen.

2.2 Site-directed mutagenesis

Site-directed mutations were introduced into genes cloned in pGEM-11Zf(+) or directly into pET3a expression vector. The synthetic oligonucleotide primers (Applied Biosystems and Bonsai Technology) were designed to contain the desired mutation. Mutant construction was carried out by two different methods. In the first, the gene encoding the halophilic dehydrogenases were cloned into pGEM-11Zf(+) and site-directed mutagenesis was performed using the GeneEditorTM *in vitro* Site-Directed Mutagenesis System (Promega). This method works by the simultaneous annealing of two oligonucleotide primers to one strand of a denatured plasmid. One primer introduces the desired mutation in the gene; and the other primer mutates the beta-lactamase gene, increasing the resistance to alternate antibiotics as penicillins and cephalosporines. The last change is important to select plasmids derived from the mutant strand. This positive selection results in consistently high mutagenesis efficiencies. The protocols supplied with the kit consist in the annealing of the two oligonucleotide primers to an alkaline-denatured dsDNA template. Following hybridization, the oligonucleotides are extended with DNA polymerase to create a double-stranded structure. The nicks are then sealed with DNA ligase and the duplex structure is used to transform an *E. coli* host. The construction of the mutants was carried out following the Promega protocol but with one modification: the length of the DNA denaturation stage was increased from 5 min at room temperature to 20 min at 37 °C due to the increase of the GC content in the halophilic genomes. In the second, the mutagenesis procedure used followed the method of the Stratagene Quick Change kit, using *Pfu Turbo* DNA polymerase from Stratagene. Extension of the oligonucleotide primers generated a mutated plasmid containing staggered nicks. Following temperature cycling, the product was treated with *Dpn* I (Fermentas). The *Dpn* I endonuclease is specific for methylated and hemimethylated DNA and was used to digest the parental DNA template and to select for mutation containing synthesized DNA. The nicked vector DNA containing the desired mutations was transformed into XL1-Blue competent cells (CNB Fermentation Service). In both methods, putative mutants were screened by dideoxynucleotide sequencing with ABI3100 DNA sequencer (Applied Biosystems).

2.3 Protein preparation

Expression *E. coli* BL21(DE3) cells were transformed with the mutated plasmid. Expression, renaturation and purification of recombinant mutants were as previously described for wild type halophilic enzymes (Pire et al., 2001; Camacho et al., 2002). The purity of the proteins was checked by SDS–polyacrylamide gel electrophoresis (SDS-PAGE). No protein contamination was detectable after Coomassie-blue staining of the gel. Protein concentration was determined by the method of Bradford (Bradford, 1976).

2.4 Glucose dehydrogenase analysis

2.4.1 Kinetic assays and data processing

Initial velocity studies were performed in 20 mM Tris–HCl buffer pH 8.8, containing 2 M NaCl and 25 mM $MgCl_2$. The reaction was monitored by measuring the appearance of NAD(P)H at 340 nm with a Jasco V-530 spectrophotometer. One unit of enzyme activity was defined as the amount of enzyme required to produce 1 µmol NAD(P)H/min under the assay conditions (40 °C).

The kinetic constants were obtained from at least triplicate measurements of the initial rates at varying concentrations of D-glucose and $NAD(P)^+$. Kinetic data were fitted to the sequential ordered BiBi equation with the program SigmaPlot 9.0.

2.4.2 Effect of EDTA concentration

The samples at different NaCl concentration were incubated with increasing EDTA concentration for 5 min at room temperature. After the incubation, the residual activities of the enzymes were measured in the activity buffer defined previously (Bonete et al., 1996).

2.4.3 Effect of temperature on enzymatic stability and activity

The samples at different NaCl concentration were incubated at various temperatures: 55, 60, 65, 70 and 80 °C. Aliquots were withdrawn at given times for measurement of residual activity. Furthermore, enzymatic activity was assayed between 25 and 75 °C at the same conditions described previously.

2.4.4 Effect of salt concentration on enzymatic activity and stability

The enzymatic activity was measured, as previously described, in buffer with KCl or NaCl in the concentration range of 0-4 M. The results are expressed as the percentage of the activity relative to the highest activity obtained.

Salt concentration stability studies were carried out at room temperature and at 40 °C. Purified preparations of enzyme in 2 M KCl were quickly diluted with 50 mM potassium phosphate buffer pH 7.3 to obtain 0.25 and 0.5 M KCl concentrations. Samples were removed at known time intervals, cooled on ice, and the residual enzymatic activity was then measured. The results are expressed as the percentage of the activity relative to that existing before incubation.

2.4.5 Differential scanning calorimetry (DSC)

DSC experiments were performed using a VP-DSC microcalorimeter (MicroCal). Temperatures from 40 °C to 90 °C were scanned at a rate of 60 °C/h using 50 mM potassium phosphate buffer pH 7.3 containing 1 mM EDTA and 0.5 M or 2.0 M KCl, which also served for baseline measurements. Prior to scanning, all samples of protein and buffer were degassed under vacuum using a ThermoVac unit (MicroCal). The protein concentrations were in the range of 50–80 μM (approximately 4–6 mg/ml). The data were analyzed using ORIGIN software v 7.0.

2.5 Isocitrate dehydrogenase analysis

2.5.1 Sequence alignment

Initial alignment with *Hfx. volcanii* NADP-dependent ICDH (Q8X277) was obtained with ClustalW (Thompson et al., 1994), taking account of information of *Bacillus subtilis* (P39126) (Singh et al., 2001) and *E. coli* (P08200) (Hurley et al., 1991) NADP-dependent ICDH, and *Thermus thermophilus* NAD-dependent IMDH (P00351) (Imada et al., 1991) and their sequences. The crystalline structures of all of them were solved previously by high-resolution X-ray analysis. Residues critical to substrate binding were identified from high-resolution crystallographic structures of *E. coli* NADP–ICDH with bound isocitrate (Hurley et al., 1991). Critical residues for coenzyme specificity were identified from high-resolution X-ray crystallographic structures of *E. coli* ICDH complexed with NADP+ (Hurley et al., 1991) and *T. thermophilus* IMDH complexed with NAD+ (Hurley & Dean, 1994).

Oligonucleotide primers containing the necessary mismatches were used for construction of the mutations: R291S, K343D, Y344I, V350A and Y390P.

2.5.2 Kinetic assays and data processing

The activities of native and mutant ICDHs were determined spectrophotometrically at A_{340} and 30 °C in 20 mM Tris-HCl buffer pH 8.0, 1 mM EDTA, 10 mM $MgCl_2$ (Tris/EDTA/Mg^{2+}) containing 2 M NaCl, 1 mM D,L-isocitrate (Camacho et al., 1995, 2002), with NADP+ or NAD+ as the coenzyme. One unit of enzyme activity is the reduction of 1 μmol of NADP per min. Initial velocities were determined by monitoring the production of NADPH or NADH at 340 nm in a 1-cm light path, based on a molar extinction coefficient of 6200 M^{-1} cm^{-1}. Kinetic parameters K_m and V_{max} were calculated for the NADP+ and NAD+ and isocitrate, depending on the cases, and the turnover number (K_{cat}) and catalytic efficiency (K_{cat}/K_m) were determined for each of the mutants, by fitting the data to the Eadie–Hofstee equation with the SigmaPlot program (Version 1.02, Jandel Scientific, Erkath, Germany) (Rodriguez-Arnedo et al., 2005).

2.5.3 Modeling ICDH

Native ICDH and the mutant ICDH with all five amino acids substituted (SDIAP mutant) were modeled with the Swiss-Model program on ExPASy Molecular Biology Server (http://swissmodel.expasy. org/) based on sequence homology. The program uses Blast and ExNRL-3D (derived from PDB) database for the search of a potential protein mold. These proteins, previously resolved by X-ray analysis, with more than 20 amino acids in length and more than 25% sequence identity were chosen. The construction of the structural model was done with the Promodll program and the minimization of energy with Gromos96. The program calculates all levels of identity between the sample problem and the sequence pattern, and it calculates the relative standard deviation to the average of the corresponding structures models and control. *Hfx. volcanii* ICDH shares 56.6% identity with *E. coli* ICDH (Camacho et al., 2002). The final image was refined with Swiss-Pdb Viewer (Rodriguez-Arnedo et al., 2005).

3. Results

3.1 Analysis of acidic surface of *Hfx. mediterranei* GlcDH

3.1.1 Choice of the halophilic GlcDH mutations

Generally, halophilic enzymes present a characteristic amino acid composition, showing an increase in the content of acidic residues and a decrease in the content of basic residues, particularly lysines. The latter decrease appears to be responsible for a reduction in the proportion of solvent-exposed hydrophobic surface. This role was investigated by site-directed mutagenesis of GlcDH from *Hfx. mediterranei*, in which three surface aspartic residues of the 27 per subunit were changed to lysine residues. At the start of the project, an initial GlcDH structure had been solved at medium resolution. Based on direct observation of this structure, the three aspartic residues chosen were D172, D216 and D344, which at least have the carboxyl oxygens exposed to the solvent (Fig. 1).

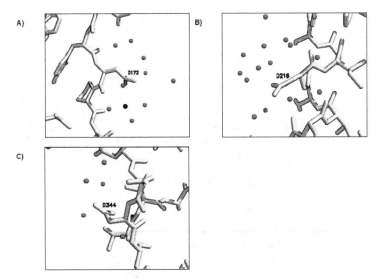

Fig. 1. Diagram showing details of the region surrounding residues D172 (A), D216 (B) and D344 (C) in the high resolution structure of D38C GlcDH with NADP$^+$ and zinc (Britton et al., 2006). The water molecules and the potassium ions are shown in red and black, respectively.

The three selected residues are considered as surface acidic residues, and they are located in different regions of the protein surface. Later, the 1.6 Å resolution GlcDH structure revealed that the side-chain carboxyl of D172 is involved in interactions with a cluster of surface water molecules near a bound potassium counter-ion. In contrast, the side-chain carboxyl of D216 forms interactions with surface waters in a region in which no counter-ions can be seen. The side-chain carboxyl of D344 lies on the surface, where it interacts with the solvent but also makes hydrogen bonds to the nearby side-chains of T346 and T347. Moreover, multiple alignments (data not shown) with other GlcDH sequences belonging to the MDR superfamily have shown that the acidic residue D216 from *Hfx. mediterranei* GlcDH is conserved in all other halophilic microorganisms. However, residue D344 is only conserved in the *Hfx. volcanii* GlcDH; and residue D172 is not present in any halophilic GlcDH. At the locations corresponding to D172, D216 and D344 in wild type *Hfx. mediterranei* GlcDH, there are non-acidic residues in 100% of the non-halophilic GlcDH sequences analyzed. Therefore, the presence of these acidic residues in the GlcDH from *Hfx. mediterranei* could be an adaptive response to the halophilic environment (Esclapez et al., 2007).

3.1.2 Site-directed mutagenesis and expression of the mutant proteins

Four mutant enzymes were obtained, the triple mutant and the three corresponding single mutant. The triple mutant GlcDH was created with the GeneEditorTM *in vitro* Site-Directed Mutagenesis System (Promega) by introducing the mutations one by one. The single mutant D172K was achieved as the first step in the constructions of the triple mutant GlcDH. The mutants D216K and D344K were constructed by PCR using *Pfu Turbo* DNA polymerase and following digestion with the endonuclease *Dpn*I.

The four mutant genes were cloned into the pET3a expression vector, and the resulting constructs were transformed into *E. coli* BL21(DE3). The expression assays were performed as described previously (Pire et al., 2001). The four mutant proteins were obtained as inclusion bodies, which were solubilized using 20 mM Tris–HCl buffer pH 8.0, 8 M Urea, 50 mM DTT and 2 mM EDTA, like wild type GlcDH. The refolding of each mutant protein was achieved by rapid dilution in 20 mM Tris–HCl buffer pH 7.4, 1 mM EDTA and KCl or NaCl in the concentration range of 1–3 M. The wild-type and triple mutant GlcDHs behave identically in the refolding process under the conditions assayed. The profiles for the triple mutant protein are like the wild type GlcDH, independently of concentration and type of salt. The three single mutants also presented the same profiles. In the presence of NaCl, the recovery of activity was always higher than with KCl; and the highest enzymatic activity was obtained at 3 M NaCl. Furthermore, at low salt concentrations the recovery of activities were lower than at high salt concentrations. No activity was recovered at 1 M KCl or NaCl. Thus, the mutations introduced on the protein surface did not appear to affect refolding in either the triple mutant or the single mutant proteins.

The purification of the GlcDH mutants were carried out as described previously. However, after 3–4 days, protein precipitation was observed in the fractions of triple mutant GlcDH whose protein concentration was greater than 1 mg/ml. This problem was solved by decreasing the protein concentration or by reducing the salt concentration through dialysis against the buffer containing 1 M NaCl or KCl. This fact indicates that the halophilic properties of the triple mutant protein have been altered, since the wild-type and single mutant proteins were stable for months under these conditions.

3.1.3 Properties of the mutant enzymes

The kinetic parameters of the mutant proteins were determined and compared to those that had previously been obtained with wild-type GlcDH. Their K_m values for $NADP^+$ and glucose are essentially similar and no significant differences in the values for V_{max} were detected. These results indicated that the kinetic parameters were not affected by the mutations. It is unlikely, therefore, that the mutations in position 172, 216 and 344 influenced the active site or the integrity of the enzyme. Similar results were obtained when residues on the surface were mutated on malate dehydrogenase from *Haloarcula marismortui* (Madern et al., 1995) and dihydrolipoamide dehydrogenase from *Hfx. volcanii* (Jolley et al., 1997).

The dependence of enzymatic activity on the concentration of NaCl is shown in Fig. 2. The triple mutant GlcDH shows its maximum activity in a buffer with 0.50–0.75 M NaCl while the wild-type protein has its maximum activity with 1.5 M NaCl. Furthermore, at low salt concentrations the activity of the triple mutant enzyme is higher than the activity of the wild-type GlcDH. At higher salt concentrations, it is lower than the wild-type protein. With the purpose of determining if the observed behavior in the triple mutant protein is due to the presence of just one mutation or of the three modifications, these experiments were also performed with each single mutant protein. The mutants D172K GlcDH and D216K GlcDH show the same profiles as the triple mutant enzyme. In striking contrast, the behavior of the D344K mutant protein is very similar to the profile obtained with the wild-type GlcDH. These results suggest that the D344K modification does not disturb the halophilic characteristics of GlcDH. Therefore, the behavior of the triple mutant GlcDH in the salt concentrations assayed could be due to the introduction of the mutation D172K and D216K. The profiles obtained using buffers with KCl are very similar.

At optimal salt concentration, the activities of the wild-type and mutant GlcDH proteins are very close. The kinetic parameters are very similar too. Therefore, it appears that the different mutations introduced in GlcDH only influence the dependence of enzymatic activity on the salt concentration. However, in similar studies with the dihydrolipoamide dehydrogenase from *Hfx. volcanii*, mutants with only one mutation (E243Q, E423S or E423A) resulted in enzymes less active than the wild-type enzyme and with different kinetic parameters. Based on these results, Jolley and co-workers (Jolley et al., 1997) also supported the view that it is the precise structural organization of acidic residues that is important in halophilic adaptation and not only the increase in acidic residue content (Madern et al., 1995; Irimia et al., 2003).

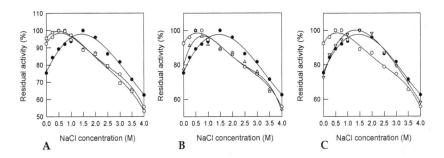

Fig. 2. Effect of NaCl on the activity of wild-type GlcDH (•), triple mutant GlcDH (○) and single mutants: (A) D172K GlcDH (□), (B) D216K GlcDH (△) and (C) D344 K GlcDH (▽). The activity buffer was 20 mM Tris–HCl pH 8.8 with varying concentrations of NaCl.

The effects of different salt concentrations on the residual activity of wild-type halophilic GlcDH and the four mutant proteins were measured after incubation at 25 °C and 40 °C. In the presence of 2 M KCl, neither wild-type enzyme nor mutant proteins were inactivated at the temperatures assayed. In particular, at salt concentrations above 1 M, the proteins were stable for weeks. As salt concentration increases, the proteins were more stable independent of the temperature. However, at low salt concentrations, small differences were observed in the stability of the proteins. The triple mutant and each single mutant protein appeared to be slightly more stable than the wild-type protein at 0.25 and 0.50 M KCl. The behavior of the proteins at 25 °C was similar, although a decrease in the temperature implies an increase of the period over which the enzymes are stable. The half-life time ($t_{1/2}$) for each protein was calculated (Table 1) showing that the mutant protein half-life times, either as a single alteration or altogether, are longer than wild type, both at 25 °C and 40 °C. However, there are no significant differences between the triple mutant and the single mutant proteins. All showed similar half-life times under the conditions assayed.

Biocalorimetry experiments were carried out under two different KCl concentrations using a DSC. In the presence of 2 M KCl, wild-type and single mutant GlcDH denaturing temperatures range from 74.6 °C to 75.9 °C. However, the triple mutant enzyme shows a lower denaturing temperature, between 73.6 °C and 73.7 °C. In other words, the triple mutant enzyme is denatured at slightly lower temperatures than are the wild-type and single mutant GlcDHs in the presence of high salt. At 0.50 M KCl (low salt), the results obtained do not reveal significant data; but the protein denaturing temperatures are lower than those obtained in the presence of high salt, independent of protein type (Fig. 3). This decrease was expected because the halophilic proteins are destabilized in low salt. Consequently the denaturing temperatures of the wild-type and mutant enzymes ranged from 59.8 °C to 60.7 °C. There were no significant differences between the temperatures.

	"Wild type"	Triple mutant	D172K	D216K	D344K
$t_{1/2}$ 40 °C (h)					
0.25 M KCl	14 ± 2	18 ± 3	23 ± 5	26 ± 8	25 ± 4
0.50 M KCl	86 ± 4	114 ± 9	95 ± 9	117 ± 8	114 ± 7
>1.0 M KCl	>170	>170	>170	>170	>170
$t_{1/2}$ 25 °C (h)					
0.25 M KCl	142 ± 22	246 ± 23	293 ± 30	219 ± 32	232 ± 30
0.50 M KCl	506 ± 50	613 ± 74	630 ± 113	660 ± 113	537 ± 75
>1.0 M KCl	>170	>170	>170	>170	>170

Table 1. Half-life times of "wild type" and mutant GlcDHs in the presence of different KCl concentrations at 40 °C (A) and 25 °C (B).

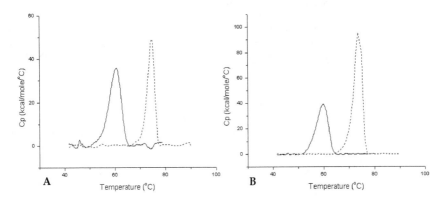

Fig. 3. Calorimetric traces of the thermal transition for wild-type GlcDH (A) and triple mutant GlcDH (B). Thermal transitions were determined in 50 mM potassium phosphate buffer pH 7.3 with 0.5 M (continuous line) or 2 M KCl (dotted line).

The data that we have presented indicate that the halophilic properties of the mutant proteins have been modified. Their enzymatic activity and kinetic parameters have been not affected by the mutations. The triple mutant and the single mutants, D172K GlcDH and D216K GlcDH, have reached their maximum activities at lower salt concentrations than wild-type GlcDH and the D344K mutant. It appears that the D344K substitution has no effect on the salt activity profile. Strikingly, in all the cases the mutant proteins were slightly more stable at low salt concentrations than was the wild-type GlcDH , although they require high salt concentration for maximum stability, like a malate dehydrogenase mutant from *Har. marismortui* (Madern et al., 1995). The biocalorimetry analyses have revealed another difference. The single mutant and the wild-type GlcDHs showed similar denaturing temperatures in the presence of 2 M KCl, while the triple mutant enzyme presented a lower denaturing temperature. Thus, more than one of our substitutions are apparently needed to significantly modify the protein's denaturing temperature at high salt concentration. Probably, these data are the result of an alteration of the hydration shell, which is required for halophilic proteins to be stable at high salt concentrations. Analysis of the high resolution GlcDH structure has shown that the size and order of the hydration shell in the halophilic enzyme is significantly greater than in non-halophilic proteins. Analyses also show that the differences in the characteristics of the molecular surface arise not only from an increase in negative surface charge, but also from the reduction in the percentage of hydrophobic surface area due to lysine side chains. Lysine residues of halophilic enzymes tend to be more buried than those of non-halophilic proteins (Britton et al., 2006).

3.2 Analysis of the zinc-binding site of GlcDH from *Hfx. mediterranei*

3.2.1 Choice of the GlcDH mutations

Whilst sequence analysis clearly identifies *Hfx. mediterranei* GlcDH as belonging to the zinc-dependent medium chain dehydrogenase/reductase family, the zinc-binding properties of the enzymes of this family are known to vary. The family includes numerous zinc-containing dehydrogenases, which bind one or two zinc atoms per subunit. One of the zinc atoms is essential for catalytic activity, while the other has a structural role and is not present in all the family members. Previous biochemical studies established that the *Hfx. mediterranei* GlcDH appears to have a single zinc atom per subunit. The role of this zinc atom is to participate in the catalytic function of the enzyme (Pire et al., 2000; Pire et al., 2001).

In the crystal structure of horse liver alcohol dehydrogenase (HLADH), three protein ligands, C46, H67 and C174 coordinate the catalytic zinc (Eklund et al., 1981). Residues analogous to C46 and H67 are conserved in the vast majority of members of the MDR family, while in some enzymes the analogous residue for C174 is glutamate as in *Thermoplasma acidophilum* GlcDH (Fig. 4). On the basis of sequence alignment, the residues involved in binding the catalytic zinc in *Hfx. mediterranei* GlcDH are predicted to be D38, H63 and E150. This sequence pattern of residues that bind the catalytic zinc has not previously been observed for any enzyme in the MDR family. The change of the C38 to D38 in the halophilic enzyme could be an adaptive response to the halophilic environment.

In order to investigate the mode of zinc binding to the halophilic GlcDH, two mutant enzymes were constructed by site-directed mutagenesis. We replaced the D38 present in the active center of the protein with C or A.

3.2.2 Site-directed mutagenesis and expression of the mutant proteins

Site-directed mutagenesis was carried out to replace the D38 residue by cysteine and alanine in the recombinant GlcDH using GeneEditorTM *in vitro* Site-Directed Mutagenesis System. The mutant genes were cloned into the pET3a expression vector. The resulting constructs were introduced by transformation into *E. coli* BL21(DE3). After expression, both mutant proteins were obtained as inclusion bodies, as was wild-type GlcDH.

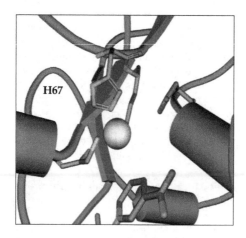

Fig. 4. The catalytic zinc-binding site in *Thermoplasma acidophilum* GlcDH.

The mutant enzymes were refolded and purified as described previously (Pire et al., 2001). In both mutants, the activity was lower than that of the wild-type protein, with the D38A mutant being inactive. This result suggests that D38 is an important residue and that the mutation to A38 leaves the enzyme seriously compromised. With respect to the D38C mutant, the maximum activity observed was approximately 30% of the activity of the wild-type enzyme.

3.2.3 Characterization of D38C GlcDH

The kinetic parameter values for mutant D38C GlcDH were determined and compared with those obtained for wild-type GlcDH (Table 2). K_{mNADP^+} differences are not significant; however, the mutation led to a significant increase of the K_m for glucose. Moreover, as the K_{cat} and $K_{cat}/K_{mglucose}$ parameters show, the catalytic efficiency of the mutant protein is less than the catalytic efficiency of wild-type GlcDH. These results indicate that the replacement of D38 to C38 in the GlcDH probably affects not only the catalytic zinc-binding site, but also the active site of the protein. The C38 GlcDH decreases the enzyme's affinity for glucose and its V_{max} relative to the wild-type enzyme. Consequently, the catalytic efficiency of the mutant enzyme is reduced.

	K_{mNADP^+} (mM)	$K_{mglucose}$ (mM)	V_{max} (U/mg)	K_{cat} (min^{-1})	K_{cat}/K_m (mM^{-1}min^{-1})
"Wild type"	0.035 ± 0.004	2.8 ± 0.3	397 ± 15	31 ± 1	11.10 ± 1.6
D38C	0.044 ± 0.010	12.4 ± 2.3	83 ± 9	7 ± 1	0.52 ± 0.14

Table 2. Kinetic parameters of recombinant wild-type GlcDH and the D38C mutant.

The zinc ion in the wild-type enzyme can be removed by EDTA treatment to yield an inactive enzyme (Pire et al., 2000). In order to compare the strength of zinc binding in "wild type" and in the D38C mutant, a similar treatment was carried out. Fig. 5 shows that zinc is more weakly bound in the D38C mutant than in the wild-type enzyme. The EDTA concentration needed to inactivate the enzyme is lower than that needed for the wild-type enzyme, and this inactivation was independent of salt concentration. For the wild-type enzyme, the capacity of EDTA to sequester the zinc is lower in the D38C mutant; and it is salt concentration-dependent. In the three NaCl concentration tested, the enzyme lost approximately 80% of its activity in the presence of 0.25 mM EDTA, and it was completely inactive at concentrations higher than 2 mM. However, in the case of the wild-type GlcDH, the EDTA necessary to sequester zinc atom at 3 M NaCl is higher than at 1 M, so the behavior of this protein is dependent on the salt concentration. At concentrations above 4 mM of the chelating agent, the enzyme is completely inactive, regardless of the NaCl concentration. Therefore, the substitution of D38 by C38 in the protein has weakened the binding of zinc ion. The D residue at position 38 in the halophilic glucose dehydrogenase instead of C, which is commonly found at the analagous postion in other members of the medium chain dehydrogenase family, could represent a halophilic adaptation.

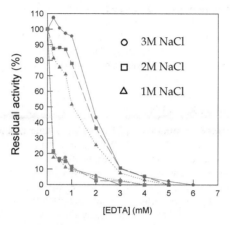

Fig. 5. Deactivation of the wild-type (blue) and D38C mutant (red) GlcDH under various EDTA concentrations at different buffer salt concentrations.

The replacement of D38 by C38 makes the binding of catalytic zinc ion of the halophilic GlcDH very similar to that presented by the thermophilic GlcDHs and other MDR family proteins. In order to clarify if C38 instead of D38 modifies the thermal characteristics of the enzyme at different salt concentrations, the effect of the temperature on enzymatic stability and activity were determined.

Generally at low salt concentration, halophilic proteins are less stable. High temperatures can contribute to their destabilization under these conditions. At high salt concentrations, halophilic proteins are stable; but stability can be perturbed by several factors, such as high temperatures. The thermal inactivation results illustrate that both the wild-type and the D38C mutant proteins show higher thermostability when the concentration of NaCl is raised. However, the D38C GlcDH appears to be slightly more thermostable than "wild type" GlcDH at the NaCl concentration assayed. The half-lives calculated for each protein under the different conditions are shown at Table 3. In general, at temperatures of 60-70 °C the D38C mutant shows a half-life higher than that of wild-type GlcDH. No reliable comparisons can be made at 80 °C, as at that temperature total inactivation of the enzyme is achieved in a few seconds. Below 60 °C the differences between the half-lives are not significant.

| | $t_{1/2}$ 1 M NaCl (h) | | $t_{1/2}$ 2 M NaCl (h) | | $t_{1/2}$ 3 M NaCl (h) | |
	D38C	"Wild type"	D38C	"Wild type"	D38C	"Wild type"
55 °C	33.9	37.2	(a)	(a)	(a)	(a)
60 °C	7.4	4.6	123.7	51.33	210.6	96.27
65 °C	0.3	0.3	9.6	8.3	(a)	(a)
70 °C	(b)	(b)	0.3	0.2	17.2	8.25
80 °C	(b)	(b)	0.04	0.02	0.2	0.2

(a) The enzyme is stable under these conditions.
(b) The enzyme is stable for only a few minutes under these conditions.

Table 3. Half-life time at different temperatures and salt concentrations of wild-type GlcDH and D38C GlcDH.

The replacement of D38 by C38 appears to have a stabilizing effect on the ability of the protein to withstand high temperatures, producing an enzyme that is marginally more stable at high temperature. However, it is clear that the enzymatic activity of the mutant is lower.

3.3 Alteration of coenzyme specificity in *Hfx. mediterranei* GlcDH and *Hfx. volcanii* ICDH

3.3.1 *Hfx. mediterranei* GlcDH

3.3.1.1 Mutations for the reversal of coenzyme specificity

The ability of dehydrogenases to discriminate between NAD⁺ and NADP⁺ lies in the amino acid sequence of the nucleotide-binding βαβ motif. This βαβ motif is centered around a highly conserved Gly–X–Gly–X–X–Gly sequence (where X is any amino acid) connecting the first β strand to the α helix. The presence of an aspartic residue at the C-terminal end of the second β strand is conserved in NAD⁺-specific enzymes. In many NADP⁺-specific enzymes, this residue is replaced by a smaller and neutral residue and complemented by a nearby positively charged residue that forms a positively charged binding pocket for adenosine 2'-phosphate. The three-dimensional structure of the cofactor binding-site of *Hfx. mediterranei* GlcDH (Fig. 6) indicates the spatial location of the residues mutated here and the interaction of R207 and R208 with the 2'-phosphate group of NADP⁺ (Britton et al., 2006; Pire et al., 2009).

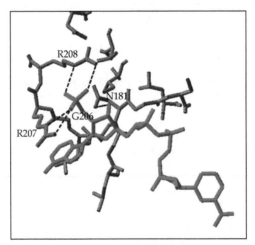

Fig. 6. View of NADP⁺ bound to "wild type" GlcDH. Interaction through hydrogen bonds is represented with dotted lines. NADP⁺ is shown in red; a portion of GlcDH, in gray. Crystal structure of GlcDH is from *Hfx. mediterranei* (PDB code 2B5V, Britton et al., 2006).

3.3.1.2 Protein properties

All the reversal coenzyme specificity mutants were expressed as inclusion bodies, and refolding was carried out by rapid dilution in the same way as for the wild-type enzyme

(Pire et al., 2001). To assess that the enzymes reached their maximum activity in terms of proper refolding, enzyme activity was measured as a function of time after rapid dilution. The wild-type and mutated enzymes behaved similarly during refolding, although the refolding kinetics of the mutants were slower. Maximum activity was reached after approximately 24 h with the mutated enzymes, whereas the wild-type enzyme achieved maximum activity 2 h after the rapid dilution of solubilized inclusion bodies (Pire et al, 2009).

Once the protein was folded, the purification procedures were identical for the wild-type and mutant enzymes (Pire et al., 2001).

3.3.1.3 Kinetics of "wild type" and coenzyme specificity reversal mutant enzymes

The kinetic constants of the wild-type and mutant forms of GlcDH were determined with both coenzymes, NAD$^+$ and NADP$^+$. The kinetic constants for the enzymes are compared in Table 4 A and B.

The K_m value of the wild-type enzyme was 11-fold lower for NADP$^+$ than for NAD$^+$, indicating that the enzyme has a strong preference for NADP$^+$. The single substitution G206D increased the K_m 74-fold for NADP$^+$ and decreased K_{cat} 2-fold, resulting in a 150-fold decrease in the K_{cat}/K_m when using NADP$^+$. This was to be expected as the negative charge of D206 would be likely to repel the adenosine 2'-phosphate of NADP$^+$. This single substitution had a positive effect on catalysis with NAD$^+$. In NAD$^+$-dependent enzymes, an aspartic residue in this position confers specificity towards NAD$^+$ by the bidentate hydrogen bonding with the 2' and 3' hydroxyl groups of the adenosine of NAD$^+$. The K_m in the presence of NAD$^+$ was similar to that of the "wild type", but K_{cat} showed a 2-fold increase. The G206D mutant preferred NAD$^+$ to NADP$^+$, showing a K_{cat} value with NAD$^+$ similar to that of the wild-type enzyme with NADP$^+$; however, the K_{cat}/K_m ratio was still better in the wild-type enzyme with NADP$^+$.

The single mutant R207I showed an increase of 48 times in K_m value with NADP$^+$ when compared with the "wild type"; this again was as expected, considering the role of D207 in the stabilization of the negative charge of the adenosine 2'-phosphate group of NADP$^+$. This increase was accompanied by a decrease in K_{cat}, which clearly makes the R207I mutant less efficient in catalysis with NADP$^+$. For NAD$^+$ the K_m value also increased, but at a ratio of 3 times, much lower than the K_m increase with NADP$^+$. The R207I substitution also makes the enzyme less efficient with NAD$^+$, with a decrease of 4 times in K_{cat}/K_m; this substitution also increases the K_m for glucose. A similar effect was even more pronounced in the single substitution R208N, in which saturation with glucose cannot be achieved, and attempts to calculate K_m and K_{cat} with both coenzymes led to very high standard deviation values.

The activity of the G206D/R207I double mutant with NADP$^+$ was very low (almost undetectable), and as such the kinetic parameters could not be calculated. However, when the coenzyme NAD$^+$ was incubated with this double mutant, it reached the highest K_{cat} value, between 1.5 and 2 times higher than the K_{cat} of the wild-type enzyme with NADP$^+$, and between 3 and 4 times higher than the K_{cat} of the "wild type" with NAD$^+$. These values indicate that the local rearrangement of the active centre due to the mutations makes catalysis more efficient. The dissociation constant for NAD$^+$ in the double mutant decreased 1.7-fold in comparison with K_{iNAD+} in the "wild type", but the K_{mNAD+} value of NAD$^+$ registered a 2-fold increase. The G206D/R207I/R208N triple substitution produced an

A)	NADP+ Wild type	G206D	R207I		
$K_{iNAD(P)}^+$ (mM)	0.09 ± 0.02	0.69 ± 0.10	1.3 ± 0.5		
$K_{mNAD(P)}^+$ (mM)	0.035 ± 0.009	2.6 ± 0.4	1.7 ± 0.3		
$K_{mglucose}$ (mM)	2.8 ± 0.3	52 ± 8	24 ± 8		
ᵃK_{cat} (min⁻¹) (x10⁻³)	31.1 ± 1.2	15.7 ± 1.5	11.8 ± 1.5		
$K_{cat}/K_{mNAD(P)}^+$ (mM⁻¹min⁻¹) (x10⁻³)	900 ± 30	6.0 ± 1.5	7 ± 2		
$K_{cat}/K_{mglucose}$ (mM⁻¹min⁻¹) (x10⁻³)	11.1 ± 1.6	0.30 ± 0.07	0.5 ± 0.2		

B)	NAD+ Wild type	G206D	R207I	G206D/ R207I	G206D/R207I/ R208N
$K_{iNAD(P)}^+$ (mM)	2.0 ± 0.2	1.0 ± 0.2	2.1 ± 0.9	1.17 ± 0.13	1.7 ± 0.4
$K_{mNAD(P)}^+$ (mM)	0.40 ± 0.12	0.49 ± 0.16	1.3 ± 0.4	2.7 ± 0.6	1.3 ± 0.3
$K_{mglucose}$ (mM)	12.9 ± 1.4	15 ± 4	32 ± 13	80 ± 14	57 ± 11
ᵃK_{cat} (min⁻¹) (x10⁻³)	15.8 ± 0.9	34 ± 5	4.3 ± 0.8	54 ± 7	2.8 ± 0.3
$K_{cat}/K_{mNAD(P)}^+$ (mM⁻¹min⁻¹) (x10⁻³)	44 ± 17	70 ± 30	3.3 ± 0,6	20 ± 7	2.1 ± 0.7
$K_{cat}/K_{mglucose}$ (mM⁻¹min⁻¹) (x10⁻³)	1.2 ± 0.2	2.3 ± 0.9	0.13 ± 0.08	0.7 ± 0.2	0.049 ± 0.015

Substrate concentrations range used in kinetic parameters determination: Wild-type: [glucose]= 2-20 mM, [NADP+]= 0.04-0.2 mM, [NAD+]= 0.286-1mM; G206D: [NAD+]= 0.286 -1mM, [glucose]= 2.5-20 mM, [NADP+]= 0.286-4 mM, [glucose]= 2.5-40 mM; R207I: [NAD+]= 0.8-2 mM, [glucose]= 25-100 mM, [NADP+]= 0.4-2 mM, [glucose]= 20-100 mM; G206D/R207I: [NAD+]= 0.4-4 mM, [glucose]= 10-100 mM; G206D/R207I/R208N:[NAD+]= 0.8-4 mM, [glucose]= 16.67-100 mM.
ᵃK_{cat} values are referred to the GlcDH dimer kinetic parameters and are expressed ± standard deviation.

Table 4. Kinetic constants of wild-type and mutant GlcDH.

inactive enzyme with NADP+, confirming that these two arginines are necessary for NADP+ stabilization. Regarding the kinetic parameters with NAD+, as in the double mutant G206D/R207I, the K_m values of both substrates were higher than in the "wild type"; but in the triple mutant the K_{cat} was also lower, and it was the worst catalyst.

In contrast with our results, in alcohol dehydrogenase from gastric tissues of *Rana perezi*, the complete reversal of coenzyme specificity from NADP(H) to NAD(H) was reached with the concerted mutation of three residues G223D/T224I/H225N (Rosell et al., 2003) . The single mutation G223D had no effect on catalysis with NAD+ and the double mutant G223D/T224I was a better catalyst than the single mutant with NAD+, but worse than the triple mutant. The great increase in the K_{cat} observed with the GlcDH double mutant was not observed in alcohol dehydrogenase. It appears that one or two substitutions in alcohol dehydrogenase were not sufficient to transform coenzyme specificity, whereas multiple substitutions could be effective (Rosell et al., 2003). The same, but reverse, effect was observed with NAD+-dependent xylitol dehydrogenase, in that the reverse of coenzyme specificity from NAD+ to NADP+ was achieved with the triple mutant D207A/I208R/F209S (Watanabe et al., 2005). In yeast alcohol dehydrogenase, a NAD+ specific enzyme, the D201G substitution produces an enzyme with low activity with $NADP^+$, but the G203R substitution neither affects affinity for NAD^+ or NADH nor enables reactivity with $NADP^+$, and the D201G/G203R enzyme has kinetic characteristics similar to the single D201G enzyme. R203 should be able to interact with the 2'-phosphate, but it seems that a greater change is required in the amino acid sequence to transform the specificity (Fan & Plapp, 1999).

Although there are some examples of a coenzyme specificity change from NAD^+ to $NADP^+$ with only one mutation, it seems that the specificity change from $NADP^+$ to NAD^+ is more difficult to reach with single substitutions. This study shows that G206, R207 and R208 are determinant for coenzyme specificity in *Hfx. mediterranei* GlcDH. The substitution G206D hampered the binding of $NADP^+$ and increased by a factor of two the activity with NAD^+, resulting in an enzyme that preferred NAD^+ over $NADP^+$. Double mutation G206D/R207I was enough to make an unproductive enzyme with $NADP^+$, although the more important findings were that in double mutant G206D/R207I, the specific activity of the enzyme with NAD^+ was almost twice than in the wild type with $NADP^+$, though the K_{cat}/K_m ratio was low due to the increase of K_m. In this sense, as some authors point out, we have to be cautious to interpret the K_{cat}/K_m ratio as catalytic efficiency, since at certain substrate concentrations the wild-type GlcDH catalyzes the oxidation of glucose, using $NADP^+$ as a coenzyme at a lower rate than double mutant in the presence of NAD^+ (Pire et al., 2009)

3.3.2 *Hfx. volcanii* ICDH

One of the most interesting features of proteins is the fact that they keep in their amino acid sequences a substantial record of their evolutionary histories. Surprisingly, homologous proteins in organisms that diverged billions of years are still similar enough to recognize a correspondence in the organization of conserved and variable regions. They can even be used as markers of the evolutionary process itself. Such comparisons have been performed using protein sequence alignments obtained with different algorithms. The alignments may reveal amino acids with a common origin and/or having similar positions in the corresponding three-dimensional structures of each protein.

Molecular evolution is based on the use of alignments to reconstruct gene trees representing, as closely as possible, the historic process of sequence divergence. This reconstruction requires the development of statistical models able to reproduce the process of mutation, drift and selection.

If one represents the structural alignment of a family of proteins in the form of linear sequence of amino acids, one can see that the spatial correspondence of identical amino acids is reflected in the form of sequence identity. The presence of insertions and deletions of specific parts of structure is revealed in the form of holes or gaps (Gómez-Moreno & Sancho, 2003).

The aim of alignment algorithms for amino acid sequences is the relative structural correspondence of residues. Comparative modeling is the extrapolation of the structure to a new amino acid sequence (model) from a known three-dimensional structure of at least one member (mould) of the same family of proteins. The obtained models contain sufficient information to permit experimental design with an acceptable degree of reliability or to allow structural comparison (Gómez-Moreno & Sancho, 2003).

One of the purposes of such an analysis is to select residues for mutation that may cause changes in some of the biochemical characteristics of the protein, such as the variation in coenzyme specificity.

3.3.2.1 Sequence alignment

The ICDHs belong to an ancient and divergent family of decarboxylating dehydrogenases that includes NAD-isopropylmalate dehydrogenase (IMDH) (Dean & Golding, 1997). "This family of dehydrogenases shares a common protein fold, topologically distinct from other dehydrogenases of known structure that lacks the $\alpha\beta\alpha\beta$ binding motif characteristic of the nucleotide-binding Rossman fold (Rossman et al., 1974; Chen et al, 1995). In ICDHs the adenosine moiety of the coenzyme binds in a pocket constructed from two loops and an α-helix (Hurley et al., 1991), although the latter is substituted by a β-turn in IMDH (Imada et al., 1991; Chen et al., 1995).

Dehydrogenases discriminate between nicotinamide coenzymes through interactions established between the protein and the 2'-phosphate of NADP+ and the 2'- and 3'-hydroxyls of NAD+ (Chen et al., 1995). In the NAD-binding site, the introduction of positively charged residues changes the preference of an NAD-dependent enzyme to neutralize the negatively charged 2'-phosphate of NADP+, as it has been demonstrated with engineered dihydrolipomide and malate dehydrogenases (Bocanegra et al., 1993; Nishiyama et al., 1993).

Specificity in *E. coli* ICDH is conferred by interactions among R395, K344, Y345, Y391 and R292' with the 2'-phosphate of bound NADP+ (Hurley et al., 1991; Dean & Golding, 1997). These residues are conserved in oligomeric NADP-dependent ICDHs and some monomeric NADP–ICDHs from prokaryotes. They are replaced with a variety of residues in the NAD-dependent ICDHs (Lloyd & Weitzman, 1988; Steen et al., 1997; Yasutake et al., 2003). Except for the substitution of Y391 with glutamine in *Aeropyrum pernix*, these residues are conserved in the archaeal NADP-dependent ICDHs and are in accordance with the cofactor specificity (Steen et al., 2002). However, K344 and Y345, which interact with NADP+ in ICDH, are substituted by D278 and I279 in IMDH. This enzyme preferentially uses NAD+ as a coenzyme; and D278 hydrogen bonds to the 2'-hydroxyl group of NAD+, while repelling the 2'-phosphate of NADP+ (Chen et al., 1996; Hurley et al., 1996). Thus, this residue is a major determinant of coenzyme specificity toward NAD+ and is strictly conserved in all known IMDHs (Chen et al., 1996). The specificity in IMDH is conferred by the strictly conserved D278 (D344 in ICDH numbering), which forms a double hydrogen bond with the 2'- and 3'-hydroxyls of the adenosine ribose of NAD+ (Dean & Golding, 1997). Not only are these movements incompatible with the strong 2'-phosphate interactions seen in NADP-ICDH, but the negative charge on D344 may also repel NADP+ (Hurley et al., 1996; Rodriguez-Arnedo et al., 2005).

Specificity is governed by (1) residues that interact directly with the unique 2'-hydroxyl and phosphate groups of NAD+ and NADP+, respectively; (2) more distant residues that modulate the effects of the first group; and (3) remote residues (Hurley et al., 1996). The first group of residues includes L344, Y345, Y391 and R395 (*E. coli* ICDH numbering), and it is easy to imagine that changes at these residues destroy the 2'-phosphate binding site (Hurley et al., 1996). The second group of residues includes V351. The adenine ring of NAD+ approaches the C_α of residue 351 as a result of a conformational change in the adenosine ribose of NAD+, which also brings it close to D344, suggesting an important role for the ring shift in specificity. The V351A mutant was designed to avoid obstructing this ring shift. A second key role for V351A is to accommodate the correct packing of the introduced I345 (Hurley et al., 1996). Sequence alignment of *Hfx. volcanii*, *E. coli* and *B. subtilis* NADP-dependent ICDHs with *T. thermophilus* NAD-dependent IMDH (Fig. 7) showed that amino acid residues involved in NADP+ binding are conserved in the NADP-dependent ICDHs and are located in a characteristic binding pocket constructed from two loops and an α-helix. In IMDH from *T. thermophilus*, this pocket comprises two loops and a β-turn (Dean & Golding, 1997; Rodriguez-Arnedo et al., 2005).

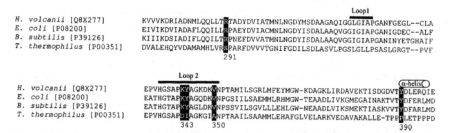

Fig. 7. Sequence alignments of *Hfx. volcanii*, *E. coli* and *B. subtilis* NADP-dependent isocitrate dehydrogenases (ICDHs) and *T. thermophilus* NAD-dependent isopropylmalate dehydrogenase. Amino acids critical to coenzyme binding and catalysis are enclosed in black boxes, and numbers correspond to relative position in the ICDH sequence from *Hfx. volcanii*.

3.3.2.2 Site-directed mutagenesis

In the halophilic enzyme, the R291S, K343D, Y344I, V350A and Y390P (halophilic ICDH numbering) mutations were selected based on homology. The substitutions were made by site-directed mutagenesis. The changes carried out are positively charged residues, such as Arg and Lys; uncharged amino acids, such as Ser; or negatively charged, such as Asp. Lys is a residue that appears to be conserved in many species, which could mean that its positive charge is crucial for proper catalysis by the enzyme. The first mutant made and characterized was R291S. Arg forms a hydrogen bond with the 2′-phosphate of a NADP+, as can be seen in the *E. coli* ICDH structure, and is replaced by Ser in *T. thermophilus* IMDH and by a wide variety of amino acids in other NAD-dependent enzymes (Chen et al., 1995). We found that, after the R291S substitution in *Hfx. volcanii* ICDH, the specificity for NADP+ decreased; but no activity for NAD+ was detected. The double mutant was obtained with a second mutation at residue Y390 in the halophilic enzyme (Rodriguez-Arnedo et al., 2005). This amino acid was replaced with Pro, as in IMDH from *T. thermophilus* (Chen et al., 1995), in order to remove a hydrogen bond to the 2′-phosphate and alter the local secondary structure from an α-helix to a β-turn. When examined, however, the secondary structure was unchanged by this substitution. Furthermore, preliminary studies (Chen et al., 1995) have shown that it is unnecessary to replace an α-helix by a β-turn to eliminate interactions with the 2′-phosphate of NADP+. Mutants three (SDP mutant) and four (SDIP mutant) involved replacing K343 and Y344 with Asp and Ile, respectively, to obtain triple and quadruple mutants. Both D343 (D278 in IMDH numbering) and I344 (I279 in IMDH numbering) are found in all known prokaryotic NAD-dependent decarboxylating dehydrogenases (Chen et al., 1996). In *Pyrococcus furiosus* NAD-dependent ICDH, the introduction of the double mutation of D328L/I329Y did not greatly decrease the preference for NAD$^+$; but it improved the preference for NADP$^+$ by reducing the catalytic efficiency for NAD$^+$ (Steen et al., 2002). In comparison, the loss of hydrogen bonding and the introduction of a negatively charged Asp greatly reduced binding with NADP$^+$, but the changes had no effect on binding with NAD$^+$ in *Hfx. volcanii* ICDH. The final mutation was V350A to obtain a quintuple mutant (SDIAP mutant) (Rodriguez-Arnedo et al., 2005).

After all five amino acids were changed in *Hfx. volcanii* ICDH, coenzyme binding switched from NADP$^+$ to NAD$^+$. The importance of V350A could explain why the quadruple mutation (SDIP mutant) destroyed NADP$^+$ binding without promoting NAD$^+$ binding. The corresponding residue is Val or Ile in all known NADP-dependent ICDHs and Ala in most, but not all, NAD-dependent dihydrodiol dehydrogenases (Hurley et al., 1996). Coenzyme specificity was changed to NAD$^+$ only after the fifth mutation, V350A. When V350A was the only change (single mutant), the coenzyme specificity was not switched to NAD$^+$, although specificity for NADP$^+$ decreased, as occurred when R291S was the only change (Rodriguez-Arnedo et al., 2005).

Molecular models of both the native enzyme and the mutants showed no significant changes in secondary structure when they were compared with the model of *E. coli* ICDH, and with each other (data not shown). The high sequence identity (56.6%) of halophilic organism isocitrate dehydrogenase with that of *E. coli* allows a reasonable model which would present a deviation of 1 Å.

The introduced mutations in ICDH from *Hfx. volcanii* do not produce apparent changes in the structure of the enzyme. Therefore, the change of coenzyme specificity would have occurred by the elimination of the hydrogen bonds formed by the amino acids involved in the binding of NADP$^+$. The position of the NADP$^+$ in the mutant is the same as it is in the native enzyme. The loop formed to accommodate the coenzyme is the same as the one seen in the *E. coli* enzyme. This loop has also been observed in the three-dimensional structure of ICDH from *B. subtilis* (Singh et al., 2001).

3.3.2.3 Kinetic characterization of mutants

Five amino acid substitutions introduced into wild-type *Hfx. volcanii* ICDH caused a complete shift in preference from NADP$^+$ to NAD$^+$. All mutations were based on sequence homology with NAD-dependent enzymes. Wild-type ICDH showed total dependence on NADP$^+$ and had a Michaelis constant (K_m) of 101 ± 30 µM and a K_{cat} of 0.176 ± 0.020 min^{-1} (Table 5). The affinity for the coenzyme was not changed until all five mutations were made, but specificity for NADP$^+$ decreased after the first mutation. Our SDIAP mutant displayed a preference for NAD$^+$ over NADP$^+$, had a K_m for NAD$^+$ of 144 ± 60 µM, and a K_{cat} of 0.0422 ± 0.0017 min^{-1} (Table 5). The K_m for NADP$^+$ increased approximately threefold in the single mutant and remained constant in the three subsequent mutants. The K_{cat}/K_m for the NADP$^+$ coenzyme was reduced 24-fold in the quadruple mutant. In contrast to the results obtained with other NAD(P)-dependent dehydrogenases (Nishiyama et al., 1993; Chen et al., 1997; Steen et al., 2002), our experiments showed that none of the mutants of *Hfx. volcanii* ICDH displayed dual cofactor specificity. Table 5 indicates that the K_{cat} for NAD$^+$ is lower than the K_{cat} for NADP$^+$ for the wild-type and some mutants, suggesting that most of the changes in specificity arise from discrimination in binding, rather than direct changes in catalysis. Catalytic efficiency of the quintuple mutant (SDIAP) with NAD+ was 17% of that of the wild-type enzyme with NADP+, suggesting that a hydrogen bond between the adenosine ribose of NAD+ and D343, as seen in the X-ray structure of the IMDH binary complex (Dean & Golding, 1997), may have been successfully established in the halophilic enzyme purified from *E. coli*. Thus, we have obtained an ICDH mutant with clearly changed coenzyme specificity (Rodriguez-Arnedo et al., 2005).

Isocitrate specificity changed with the first mutation, but in this case, specificity for isocitrate increased 3- to 10-fold with increasing mutations, suggesting that these mutations favored substrate binding. The maximum value for isocitrate binding occurred when the mutant showed specificity for NAD+ only. Thus, the mutations markedly influenced not only the K_m for NAD(P)+, but also the K_m for isocitrate. The effect of the mutation on the efficiency for NADP+ or NAD+ was evaluated by incorporating the K_m for the substrate. This parameter is called the overall catalytic efficiency and is defined as: $(K_{cat}/(K_m^{IC} \times K_m^{NAD(P)})$ (Table 5) (Nishiyama et al., 1993). We speculate that some specific interaction between the substrate and NADP+, which differs from the native substrate–coenzyme complex, is responsible for the decrease in activity. The ratio K_{cat}/K_m is a measure of both enzyme efficiency and the degree to which an enzyme stabilizes the transition state (Dean & Golding, 1997).

Residues at 291 343 344 350 390	NADP+ K_m (μM)	K_{cat} (min-1)	K_{cat}/K_m x 10³ (min-1/μM)	NAD+ K_m (μM)	K_{cat} (min-1)	K_{cat}/K_m x 10³ (min-1/μM)
R K Y V Y	101±30	0.176±0.020	1.7±0.7			
S - - - -	350±30	0.244±0.020	0.70±0.12			
S - - - P	327±50	0.060±0.004	0.18±0.04			
S D - - P	203±30	0.145±0.008	0.70±0.14			
S D I - P	282±20	0.020±0.003	0.071±0.016			
S D I A P	___	___	___	144±60	0.0422±0.0017	0.29±0.13

Residues at 291 343 344 350 390	Isocitrate K_m (μM)	K_{cat} (min-1)	K_{cat}/K_m x 10³ (min-1/μM)	$K_{cat}/(K_m^{IC} \times K_m^{NAD(P)})$
R K Y V Y	108±30	0.139±0.010	1.3±0.5	1.3 x 10-5
S - - - -	18±3	0.123±0.002	6.8±1.2	2.0 x 10-5
S - - - P	36±4	0.032±0.003	0.88±0.18	2.7 x 10-6
S D - - P	36±5	0.086±0.003	2.4±0.4	1.2 x 10-5
S D I - P	40±7	0.0149±0.0003	0.37±0.07	1.3 x 10-6
S D I A P	11±1*	0.0327±0.0013	3.0±0.4*	2.0 x 10-5*

Initial rates were measured spectrophotometrically at 30 °C in 20 mM Tris-HCl buffer pH 8.0, 1 mM EDTA and 10 mM MgCl₂ (Tris/EDTA/Mg²⁺) containing 2 M NaCl. Substrate and coenzyme concentrations were varied between 0.019 and 0.10 mM for D,L-isocitrate and between 0.044 and 0.4 mM for NADP+ or NAD+. Single-letter amino acid codes denote amino acid residues. Catalytic efficiency was measured as K_{cat}/K_m. Asterisks indicate values that were obtained with NAD+ as the coenzyme. Abbreviations: K_m = Michaelis constant; K_{cat} = turnover number; K_{cat}/K_m = catalytic efficiency; and $K_{cat}/(K_m^{IC} \times K_m^{NAD(P)})$ = overall catalytic efficiency.

Table 5. Kinetic parameters of purified wild-type and mutant *Hfx. volcanii* Ds-threo-isocitrate dehydrogenases (ICDHs) toward NADP+ and NAD+.

One might think that a local conformational change induced by specific binding of NADP+ or NAD+ is responsible for the variation in the behavior of the *Hfx. volcanii* enzyme versus isocitrate. This effect is probably due to less repulsion of the charges in the active site.

The comparison of the sequence with that of the *T. thermophilus* IMDH (Imada et al., 1991) reveals a consistent framework in the evolution of the substrate specificity of the decarboxylating dehydrogenases belonging to the class of isocitrate dehydrogenases. All residues (R117, R127, R151, Y158 and K228) interact with the α- and β-carboxylates and α-hydroxyls of isocitrate, which are common with isopropylmalate and malate; all residues are conserved in all known ICDHs and IMDHs. Two aspartic acid residues, D282 and D306, coordinate Mg^{2+}. They are also conserved in all available sequences. Residues that interact with the γ-carboxylate of isocitrate, S111 and N113 (*Hfx. volcanii* ICDH numbering), are only found in ICDH sequences and are not found in IMDH sequences. Common features of ICDH and IMDH are due to conserved residues that interact with the α- and β-groups and the metal ion. Difference in the substrate specificity is determined by non-conserved residues that interact with the γ-group (Hurley et al., 1991).

4. Conclusion

Site–directed mutagenesis has allowed us to (1) extend the understanding of the molecular basis of salt tolerance for halophilic adaptation, (2) analyze the role of sequence differences between thermophilic and halophilic dehydrogenases involving a ligand to the zinc ion, and (3) identify the residues implicated in coenzyme specificity.

The replacement of aspartic residues by lysine residues on the GlcDH surface have led to a modification of the halophilic properties of the mutant enzymes, D172K and D216K being the most significant mutations (Esclapez et al., 2007).

The mutation of D38, a residue that lies close to the catalytic zinc ion, to C38 or A38 led to a significant reduction in and abolition of activity, respectively. These results suggest that this residue is important in catalysis, either in forming a key aspect of the zinc-binding site or in some other process related with substrate recognition. The replacement of D38 by C38 results in the production of a less efficient enzyme with lower enzymatic activity and catalytic efficiency. Furthermore, this mutant shows slightly more thermostability. Although the D38C GlcDH is less active, it has been crystallized in the presence of several combinations of products and substrates. This fact has allowed us to describe many aspects of the mechanism of the zinc-dependent MDR superfamily (Esclapez et al., 2005; Baker et al., 2009).

Structural analysis of the GlcDH from *Hfx. mediterranei* revealed that the adenosine 2′-phosphate of NADP+ is stabilized by the side chains of R207 and R208. The first attempt to change coenzyme specificity involved making the G206D mutant. Further substitutions of uncharged residues for these residues were made to analyze their importance in NADP+ binding and to improve specificity for NAD+. The single mutants G206D and R207I were less efficient with NADP+ than the wild type, and the double and triple mutants, G206D/R207I and G206D/R207I/R208N, showed no activity with NADP+ (Pire et al., 2009).

The results obtained in our study with the halophilic ICDH and the complete switch of coenzyme specificities in IMDH from *T. thermophilus* (Imada et al., 1991) and ICDHs from *E. coli* show that coenzyme specificity in the β-decarboxylating dehydrogenases are principally determined by interactions between the nucleotides and surface amino acid residues lining the binding pockets (Rodriguez-Arnedo et al., 2005).

5. Acknowledgment

We thank Dr. Rice, Dr. Baker and Dr. Britton, from The University of Sheffield (UK), for helping us to prepare GlcDH structure figures. This work was supported by Grants from Ministerio de Educación (BIO2002-03179 and BIO2005-08991-C02-01).

6. References

Bonete, M. J., Pire, C., Llorca, F. I. & Camacho, M. L. (1996) Glucose dehydrogenase from the halophilic Archaeon *Haloferax mediterranei*: enzyme purification, characterisation and N-terminal sequence. *FEBS Lett.*, 383: 227-229.

Bradford, M. M. (1976) A rapid and sensitive method for the quantitation of microgram quantities of protein utilizing the principle of protein-dye binding. *Anal. Biochem.*, 72: 248-254.

Baker, P. J., Britton, K. L., Fisher, M., Esclapez, J., Pire, C., Bonete, M. J., Ferrer, J. & Rice, D. W. (2009) Active site dynamics in the zinc-dependent medium chain alcohol dehydrogenase superfamily. *Proc. Natl. Acad. Sci. USA*, 106: 779-784.

Bocanegra, J. A., Scrutton, N. S. & Perham, R. N. (1993) Creation of an NADP-dependent pyruvate dehydrogenase multienzyme complex by protein engineering. *Biochemistry*, 32: 2737-2740.

Britton, K. L., Stillman, T. J., Yip, K. S. P., Forterre, P., Engel, P. C. & Rice, D. W. (1998) Insights into the molecular basis of salt tolerance from the study of glutamate dehydrogenase from *Halobacterium salinarum*. *J. Biol. Chem.*, 273: 9023-9030.

Britton, K. L., Baker, P. J., Fisher, M., Ruzheinikov, S., Gilmour, D. J., Bonete, M. J., Ferrer, J., Pire, C., Esclapez, J. & Rice, D. W. (2006) Analysis of protein solvent interactions in glucose dehydrogenase from the extreme halophile *Haloferax mediterranei*. *Proc. Natl. Acad. Sci. USA*, 103: 4846-4851.

Camacho, M. L., Brown, R. A., Bonete, M.-J., Danson, M. J. & Hough, D. W. (1995) Isocitrate dehydrogenases from *Haloferax volcanii* and *Sulfolobus solfataricus*: enzyme purification, characterisation and N-terminal sequence. *FEMS Microbiol. Lett.*, 134: 85-90.

Camacho, M. L., Rodríguez-Arnedo, A. & Bonete, M. J. (2002) NADP-dependent isocitrate dehydrogenase from the halophilic archaeon *Haloferax volcanii*: cloning, sequence determination and overexpression in *Escherichia coli*. *FEMS Microbiol. Lett.*, 209: 155-160.

Chen, R., Greer, A. & Dean, A. M. (1995) A highly active decarboxylating dehydrogenase with rationally inverted coenzyme specificity. *Proc. Natl. Acad. Sci. USA*, 92: 11666-11670.

Chen, R., Greer, A. & Dean, A. M. (1996) Redesigning secondary structure to invert coenzyme specificity in isopropylmalate dehydrogenase. *Proc. Natl. Acad. Sci. USA*, 93: 12171-12176.

Chen, R., Greer, A. & Dean, A. M. (1997) Structural constraints in protein engineering: the coenzyme specificity of *Escherichia coli* isocitrate dehydrogenase. *Eur. J. Biochem.*, 250: 578-582.

Dean, A. M. & Golding, G. B. (1997) Protein engineering reveals ancient adaptive replacements in isocitrate dehydrogenase. *Proc. Natl. Acad. Sci. USA*, 94: 3104-3109.

Eisenberg, H., Mevarech, M. & Zaccai, G. (1992) Biochemical, structural and molecular genetic aspects of halophilism. *Adv. Protein Chem.*, 43: 1-62.

Eklund, H., Samma, J. P., Wallen, L., Brändén, C. L., Akeson, A. & Jones, T. A. (1981) Structure of a triclinic ternary complex of horse liver alcohol dehydrogenase at 2,9 Å resolution. *J. Mol. Biol.*, 146: 561-587.

Esclapez, J., Britton, K. L., Baker, P. J., Fisher, M., Pire, C., Ferrer, J., Bonete, M. J. & Rice, D. W. (2005) Crystallization and preliminary X-ray analysis of binary and ternary complexes of *Haloferax mediterranei* glucose dehydrogenase. *Acta Cryst. F*, 61: 743-746.

Esclapez, J., Pire, C., Bautista, V., Martínez-Espinosa, R. M., Ferrer, J. & Bonete, M. J. (2007) Analysis of acidic surface of *Haloferax mediterranei* glucose dehydrogenase by site-directed mutagenesis. *FEBS Lett.*, 581: 837-842.

Fan, F. & Plapp, B. V. (1999) Probing the affinity and specificity of yeast alcohol dehydrogenase I for coenzymes. *Arch. Biochem. Biophys.*, 367: 240-249.

Ferrer, J., Fisher, M., Burke, J., Sedelnikova, S. E., Baker, P. J., Gilmour, D. J., Bonete, M. J., Pire, C., Esclapez, J. & Rice, D. W. (2001) Crystallization and preliminary X-ray analysis of glucose dehydrogenase from *Haloferax mediterranei. Acta Cryst. D*, 57: 1887-1889.

Gómez-Moreno, C. & Sancho, J. (Eds) (2003). *Estructura de Proteínas*, Ariel Ciencia, ISBN 9788434480612, Barcelona, Spain.

Hurley, J. H., Dean, L. A. M., Koshland, Jr., D. E. & Stroud, R. M. (1991) Catalytic Mechanism of NADP+-Dependent Isocitrate Dehydrogenase: Implications from the Structures of Magnesium-Isocitrate and NADP+ Complexes. *Biochemistry*, 30: 8671-8678.

Hurley, J. H. & Dean, A. M. (1994) Structure of 3-isopropylmalate dehydrogenase in complex with NAD+: ligand induced loop closing and mechanism for cofactor specificity. *Structure*, 2: 1007-1016.

Hurley, J. H., Chen, R. & Dean, A. M. (1996) Determinants of Cofactor Specificity in Isocitrate Dehydrogenase: Structure of an Engineered NADP+ → NAD+ Specificity-Reversal Mutant. *Biochemistry*, 35: 5670-5678.

Imada, K., Sato, M., Tanaka, N., Katsube, Y., Matsuura, Y. & Oshima, T. (1991) Three-dimensional structure of a highly thermostable enzyme, 3-isopropylmalate dehydrogenase of *Thermus thermophilus* at 2.2 Å resolution *J. Mol. Biol.*, 222: 725-738.

Irimia, A., Ebe, C., Madern, D., Richard, S. B., Cosenza, L. W., Zaccai, G. & Vellieux, F. M. D. (2003) The oligomeric states of *Haloarcula marismortui* malate dehydrogenase are modulated by solvent components as shown by crystallographic and biochemical studies. *J. Mol. Biol.*, 326: 859-873.

Jolley, K. A., Rusell, R. J. M., Hough, D. H. & Danson, M. J. (1997) Site-directed mutagenesis and halophilicity of dihydrolipoamide dehydrogenase from the halophilic archaeon, *Haloferax volcanii. Eur. J. Biochem.*, 248: 362-368.

Kauri, T., Wallace, R. & Kushner, D. J. (1990) Nutrition of the halophilic archaebacterium, *Haloferax volcanii. Syst. Appl. Microbiol.*, 13: 14-18.

Kay, J. & Weitzman, P. D. J. (Eds.) (1987) Krebs' Citric Acid Cycle: Half a Century and Still Turning. *Biochem. Soc. Symp.*, 54: 1-198.

Lanyi, J. K. (1974) Salt-dependent properties of proteins from extremely halophilic bacteria. *Bacteriol. Rev.*, 38: 272-290.

Lloyd, A. J. & Weitzman, P. D. J. (1988) Purification and characterization of NAD-linked isocitrate dehydrogenase from *Methylophilus methylotrophus. Biochem. Soc. Trans.*, 16: 871-872.

Madern, D., Pfister, C. & Zaccai, G. (1995) Mutation at a single acidic amino acid enhances the halophilic behavior of malate dehydrogenase from *Haloarcula marismortui* in physiological salts. *Eur. J. Biochem.*, 230: 1088-1095.

Madern, D., Ebel, C. & Zaccai, G. (2000) Halophilic adaptation of enzymes. *Extremophiles*, 4: 91-98.

Madern, D., Camacho, M., Rodríguez-Arnedo, A., Bonete, M.-J. & Zaccai, G. (2004) Salt-dependent studies of NADP-dependent isocitrate dehydrogenase from the halophilic archaeon *Haloferax volcanii. Extremophiles*, 8: 377-384.

Nishiyama, M., Birktoft, J. J. & Beppu, T. (1993) Alteration of Coenzyme Specificity of Malate Dehydrogenase from *Thermus flavus* by Site-directed Mutagenesis. *J. Biol. Chem.*, 268: 4656–4660.

Pire, C., Camacho, M. L., Ferrer, J., Hough, D. W. & Bonete, M. J. (2000) NAD(P)+- glucose dehydrogenase from *Haloferax mediterranei*: kinetic mechanism and metal content. *J. Mol. Catal. B: Enzym.*, 10: 409-417.

Pire, C., Esclapez, J., Ferrer, J. & Bonete, M. J. (2001) Heterologous overexpression of glucose dehydrogenase from the halophilic archaeon *Haloferax mediterranei*, an enzyme of the medium chain dehydrogenase/reductase family. *FEMS Lett.*, 200: 221-227.

Pire, C., Esclapez, J., Díaz, S., Pérez-Pomares, F., Ferrer, J. & Bonete, M. J. (2009) Alteration of coenzyme specificity in halophilic NAD(P)+ glucose dehydrogenase by site-directed mutagenesis. *J. Mol. Catal. B: Enzym.*, 59: 261-265

Richard, S. B., Madern, D., Garcin, E. & Zaccai, G. (2000) Halophilic adaptation: novel solvent-protein interactions observed in the 2.9 and 2.6 Å resolution structures of the wild type and a mutant of malate dehydrogenase from *Haloarcula marismortui*. *Biochemistry*, 39: 992-1000.

Rodríguez-Arnedo, A., Camacho, M. & Bonete, M. J. (2005) Complete reversal of coenzyme specificity of isocitrate dehydrogenase from *Haloferax volcanii*. *Protein J.*, 24: 259-266.

Rodríguez Valera, F., Juez, G. & Kushner, D. J. (1983) *Halobacterium mediterranei* spec. nov., a new carbohydrate utilizing extreme halophile. *System. Appl. Microbiol.*, 4: 369-381.

Rosell, A., Valencia, E., Ochoa, W. F., Fita, I, Parés, J. & Farrés, X. (2003) Complete reversal of coenzyme specificity by concerted mutation of three consecutive residues in alcohol dehydrogenase. *J. Biol. Chem.*, 278: 40573-40580.

Rossmann, M. G., Moras, D. & Olsen, K. W. (1974) Chemical and biological evolution of nucleotide-binding protein. *Nature*, 250: 194-199.

Serrano, J. A., Camacho, M. & Bonete, M. J. (1998). Operation of glyoxylate cycle in halophilic archaea: presence of malate synthase and isocitrate lyase in *Haloferax volcanii*. *FEBS Lett.*, 434: 13-16.

Singh, S. K., Matsuno, K., LaPorte, D. C. & Banaszak, L. J. (2001) Crystal Structure of *Bacillus subtilis* Isocitrate Dehydrogenase at 1.55 Å. Insights Into The Nature of Substrate Specificity Exhibited by *Escherichia coli* Isocitrate Dehydrogenase Kinase/Phosphatase. *J. Biol. Chem.*, 276: 26154–26163.

Steen, I. H., Lien, T. & Birkeland, N.-K. (1997) Biochemical and phylogenetic characterization of isocitrate dehydrogenase from a hyperthermophilic archaeon, *Archaeoglobus fulgidus*. *Arch. Microbiol.*, 168: 412-420.

Steen, I. H., Lien, T., Madsen, M. S. & Birkeland, N.-K. (2002) Identification of cofactor discrimination sites in NAD-isocitrate dehydrogenase from *Pyrococcus furiosus*. *Arch. Microbiol.*, 178: 297-300.

Thompson, J. D., Higgins, D. G. & Gibson, T. J. (1994) CLUSTAL W: improving the sensitivity of progressive multiple sequence alignment through sequence weighting, positions-specific gap penalties and weight matrix choice. *Nucleic Acids Res.*, 22: 4673-4680.

Watanabe, S., Kodaki, T. & Makino, S. (2005) Complete reversal of coenzyme specificity of xylitol dehydrogenase and increase of thermostability by the introduction of structural zinc. *J. Biol. Chem.*, 280: 10340-10349.

Yasutake, Y., Watanabe, S., Yao, M., Takada, Y., Fukunaga, N. & Tanaka, I. (2003) Crystal Structure of the Monomeric Isocitrate Dehydrogenase in the Presence of NADP+. Insight into the cofactor recognition, catalysis, and evolution. *J. Biol. Chem.*, 278: 36897-36904.

Targeted Mutagenesis in the Study of the Tight Adherence (*tad*) Locus of *Aggregatibacter actinomycetemcomitans*

David H. Figurski[1], Daniel H. Fine[2], Brenda A. Perez-Cheeks[1],
Valerie W. Grosso[1], Karin E. Kram[1], Jianyuan Hua[1],
Ke Xu[1] and Jamila Hedhli[1]

[1]*Department of Microbiology & Immunology, College of Physicians & Surgeons,*
Columbia University, New York, NY, USA
[2]*Department of Oral Biology, The University of Medicine & Dentistry of New Jersey,*
Newark, NJ, USA

1. Introduction

Aggregatibacter actinomycetemcomitans is a Gram-negative, capnophilic (CO_2 loving), coccobacillus found only in humans and Old World primates (for reviews, see Henderson et al., 2003, 2010). This bacterium is primarily known as the etiologic agent of Localized Aggressive Periodontitis (LAP), which is predominantly an infection of adolescents (Slots & Ting, 1999; Zambon, 1985). *A. actinomycetemcomitans* also causes extraoral infections, including infective endocarditis, septicemia, and abscesses (Fine et al., 2006; Fives-Taylor et al., 1999; Rahamat-Langendoen et al., 2011; van Winkelhoff & Slots 1999). *A. actinomycetemcomitans* is a member of the HACEK group of Gram-negative bacteria, all of which can cause infective endocarditis (Paturel et al., 2004). Most HACEK bacteria-caused cases of infective endocarditis result from *A. actinomycetemcomitans*. Poor oral health is a risk factor for developing severe extraoral infections by *A. actinomycetemcomitans* (Paturel et al., 2004; van Winkelhoff & Slots 1999). One study showed that 16% of patients with infective endocarditis from *A. actinomycetemcomitans* had dental procedures immediately prior to the onset of disease. A staggering 42% of *A. actinomycetemcomitans*-caused infective endocarditis patients had generalized dental disease – an indication of poor overall oral health.

LAP is a severe form of periodontitis (Genco et al., 1986; Slots & Ting, 1999). For reasons so far unknown, the disease is localized to the premolar and incisor teeth. Infection leads to inflammation and rapid destruction of the periodontal ligament and the alveolar bone and culminates in loss of teeth. The prevalence of LAP has been estimated to be 0.5-1% (Henderson et al., 2002; Löe & Brown, 1991; Rylev & Kilian, 2008). However, prevalence varies considerably with different ethnic groups. For example, in the African-American population, the prevalence is 10-15 times higher than the average. There is evidence that race and socioeconomic status play key roles in determining prevalence.

The molecular mechanisms behind the pathogenesis of *A. actinomycetemcomitans* are not understood. The organism elaborates a number of factors that have been implicated in virulence (Fine et al., 2006; Fives-Taylor et al., 1999). Several adhesins have been identified. The best studied adhesins are (1) Aae, an autotransporter that binds to buccal epithelial cells (Fine et al., 2005, 2010; Rose et al., 2003); (2) ApiA, another autotransporter that binds to buccal epithelial cells, but with lower affinity than does Aae (Komatsuzawa et al., 2002; Yue et al., 2007); and (3) EmaA, which binds collagen (Jiang et al., 2012; Mintz, 2004). A leukotoxin affects leukocytes by inducing one of several pathways (Tsai et al., 1979; Kachlany, 2010). The pathway that is activated depends on the cell type. The toxic effects include apoptosis, degranulation, and a novel lysosome-mediated cell-death pathway (DiFranco et al., 2012; Fong et al., 2006; Kelk et al., 2003, 2011; Lally et al., 1999). Another factor is the cytolethal distending toxin, which causes death by cell-cycle arrest (De Rycke & Oswald, 2001; Fine et al., 2006; Pickett & Whitehouse, 1999).

1.1 Non-specific adherence and virulence

A particularly striking property of a fresh clinical isolate of *A. actinomycetemcomitans* is its ability to form an extremely tenacious biofilm on inert surfaces, such as glass, plastic, and hydroxyapatite (Fig. 1) (Fine et al., 1999a; Kachlany et al., 2000, 2001a). Freshly isolated clinical strains of *A. actinomycetemcomitans* form small, rough colonies on agar plates (Fig. 2A). In broth, the clinical isolates tightly adhere to the surfaces of culture vessels; and aggregates may be visible at the bottom of a tube. The cells express long, bundled, protein fibrils, termed "Flp fibers," which are required to form the tenacious biofilm. (A Flp fiber is composed of several Flp pili.) Propagation of adherent wild-type strains generally leads to spontaneous variants (see Section 4 below) that produce smooth colonies (Fig. 2B), have cells that do not autoaggregate, and cannot produce tenacious biofilms because the cells do not express Flp pili (Fig. 1, right) (Fine et al., 1999b; Kachlany et al., 2001a).

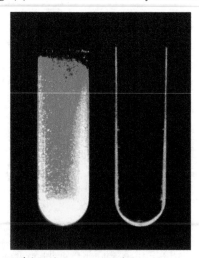

Fig. 1. Tenacious adherence of *A. actinomycetemcomitans*.
Cultures of a clinical isolate (left) and a spontaneous smooth-colony variant (right) were grown in broth. After the culture was mixed and the broth was poured out, the remaining adherent cells were stained with ethidium bromide and illuminated with UV light.

Studies in our LAP-infection model demonstrated that a wild-type strain is able to colonize and persist in the mouths of rats (Fine et al., 2001). In contrast, an isogenic smooth-colony-forming variant failed to persist. Despite this unequivocal evidence for the importance of tenacious adherence in the colonization of the oral cavity by *A. actinomycetemcomitans*, the genetic and molecular bases underlying this remarkable property were unknown.

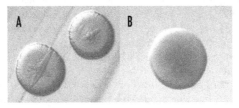

Fig. 2. Colony morphologies of *A. actinomycetemcomitans*.
(A) "Rough" colony morphology of a clinical isolate. (B) "Smooth" colony morphology of a spontaneous variant.

1.2 The *tad* locus

The study of *A. actinomycetemcomitans* had been hampered by a paucity of molecular tools for genetic manipulation, especially tools for clinical isolates. We adapted the transposon Tn903 to provide inducible random mutagenesis of the chromosome of *A. actinomycetemcomitans* (Thomson et al., 1999). A clinical isolate of *A. actinomycetemcomitans* from a 13-year-old African-American female was mutagenized using our synthetic Tn903 transposon (IS903φkan). The random-mutant bank was screened for mutant strains that formed smooth colonies. Identifying and sequencing the transposon insertion sites of smooth mutants led to the identification of a 14-gene segment, designated the *tad* (tight *ad*herence) locus (Fig. 3A) (Kachlany et al., 2000; Planet et al., 2003; Tomich et al, 2007). The *tad* locus encodes a novel macromolecular transport system that is used in the biogenesis of Flp pili (Fig. 4). One hypothetical gene of the *tad* locus, *flp-2*, is not needed for Flp pili

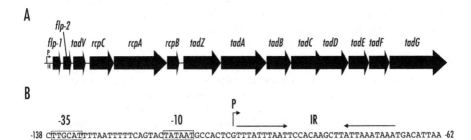

Fig. 3. The *tad* locus of *A. actinomycetemcomitans*.
(A) The size of the locus is approximately 12 kb. Filled arrows designate the genes. Their relative sizes are approximately correct. "P" and the bent arrow indicate the transcriptional start site (Haase et al., 2003); "IR," inverted repeat. (B) The promoter region magnified.

biosynthesis in *A. actinomycetemcomitans* (Perez et al., 2006). We determined whether the gene products are cytoplasmic, in the inner membrane, or in the outer membrane (Clock et al., 2008). Our studies have also revealed the functions or features of the products of several *tad* locus genes (Tomich et al., 2007). *flp-1* encodes the Flp1 prepilin (Inoue et al., 1998; Kachlany et al., 2001b); *tadV* encodes a protease that processes the Flp1 prepilin to the mature Flp1 pilin that is needed for assembly into Flp pili (Tomich et al., 2006); the product of *rcpA* (Haase et al., 1999) has the properties of an outer membrane pore (secretin) (Clock et al., 2008); the product of *rcpB* (Haase et al., 1999) is an outer-membrane protein that may gate the pore (Clock et al., 2008; Perez et al., 2006); *tadZ* encodes a protein that may localize the Tad secretion machine to a pole (Perez-Cheeks et al., 2012; Xu et al., 2012; see Section 3 below); the *tadA* product is an ATPase (Bhattacharjee et al., 2001); and the *tadE* and *tadF* products are "pseudopilins," whose functions are not known; but the pseudopilins are processed by TadV in the same way that the prepilin is (Tomich et al., 2006).

tad locus

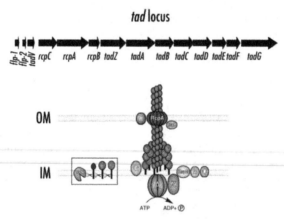

Fig. 4. Hypothetical structure of the Tad secretion apparatus.
Shown is a cartoon of the Flp pilus and its secretion machine formed (at least in part) by a complex of the *tad* gene products. The lack of detail reflects the paucity of data for the gene products. "IM" means "inner membrane"; "OM," "outer membrane."

1.3 The *tad* locus in *A. actinomycetemcomitans* is a virulence factor

Using our rat model of LAP, we demonstrated that a functional *tad* locus is needed for the ability of *A. actinomycetemcomitans* to colonize and persist in the oral cavity and to cause tissue and bone destruction (Schreiner et al., 2003). Unlike the wild-type strain, isogenic *tad* locus mutant strains, defective in either *flp-1* (encoding the pilin) or *tadA* (encoding an ATPase), did not persist in the mouths of the rats; nor did they cause bone loss. In addition to our evidence in *A. actinomycetemcomitans*, studies in other bacterial species have demonstrated the involvement of *tad* homologs in virulence: *Haemophilus ducreyi* (Spinola et al., 2003), the etiologic agent of chancroid, a sexually transmitted disease, and *Pasteurella multocida* (Fuller et al., 2000), which causes fowl cholera. The *tad* genes of *Pseudomonas aeruginosa* are needed for biofilm formation and for adherence to epithelial cells (de Bentzmann et al., 2006).

1.4 *tad* loci are widespread in prokaryotes

Searches of completed and ongoing microbial sequencing projects have revealed that closely related *tad* gene clusters are present in the genomes of a wide variety of Gram-negative and Gram-positive bacteria (Kachlany et al., 2001a; Tomich et al, 2007; P. Planet & D. Figurski unpublished results, P. Planet, unpublished results), some of which represent significant threats to human health. Among the pathogens are *Yersinia pestis*, the agent of bubonic plague; *Mycobacterium tuberculosis*, the agent of tuberculosis; *Bordetella pertussis*, the agent of whooping cough; *Burkholderia cepacia*, a frequent colonizer of the lungs of patients with cystic fibrosis; and *P. aeruginosa*, an opportunistic human pathogen. About 40% of over 3400 bacterial genomes sequenced to date have *tad* loci. In addition, potentially all Archaea may harbor homologs of *tad* genes (P. Planet, personal communication). Given our evidence that the *tad* locus is foreign to the chromosome of *A. actinomycetemcomitans* and that *tad* loci are widely distributed in prokaryotes, we have also referred to the *tad* locus as the "Widespread Colonization Island" (WCI) (Planet et al., 2003).

2. The Flp pilus

The Flp pili of *A. actinomycetemcomitans* (Fig. 5) are proteinaceous fibers that are attached to the exterior of the bacterial cell and are necessary for tenacious adherence (Kachlany et al., 2000, 2001b). Flp pili are polymers of the mature Flp1 pilin protein, and they are assembled and secreted by a complex of proteins encoded by the *tad* locus. Flp pili are abundant, extremely adhesive, and bundled. The *flp-1* and *flp-2* genes of *A. actinomycetemcomitans* and their predicted products and the *flp* genes from other organisms and their predicted products form a distinct, monophyletic group with homology to other pilin genes, particularly to those of subclass b (Kachlany et al., 2001b). We (M. Tomich and D. Figurski, unpublished results) and others (Inoue et al., 2000) have shown that the Flp1 pilin is a glycoprotein, but the structure and function of the modification is unknown.

Fig. 5. Flp pili.
Transmission electron micrograph of purified Flp pili from *A. actinomycetemcomitans*.

After their translocation to the inner membrane, the prepilins are cleaved (processed) to maturity by cognate prepilin peptidases (Giltner et al., 2012). We have shown that the 75-amino acid Flp1 prepilin of *A. actinomycetemcomitans* is cleaved by TadV protein at the sequence G^XXXXEY (Tomich et al., 2006). The mature Flp1 pilin is only 49 amino acids in length, which is much smaller than other known type IV pilins (Giltner et al., 2012; Kachlany et al., 2001b). Because of its small size, we believe Flp1 pilin is an attractive subject for genetic and structural analysis. In addition to learning about the molecular details of Flp1, we also wish to understand the basis of the three most obvious phenotypes of Flp pili:

tenacious adherence to surfaces, binding to *A. actinomycetemcomitans* cells, and binding to each other (bundling).

2.1 Alanine-scanning mutagenesis of the coding region for the mature Flp1 pilin

To begin to study the properties of Flp1, we constructed and characterized a series of Flp1 pilin mutants, each with an alanine substitution for a specific non-alanine residue of the mature Flp1 pilin. The codon for each non-alanine residue was changed to a codon for alanine. (The mutant genes were constructed with the fewest possible nucleotide changes.) In this way, translation of the mutant gene would give a mutant Flp1 prepilin, which, after being processed, gave rise to a mutant mature Flp1 pilin. (Alanine was chosen because it is the smallest amino acid that is relatively neutral and can maintain an α-helix in a polypeptide.) We changed the non-alanine residues in the mature Flp1 pilin by overlap extension PCR (polymerase chain reaction) (Ho et al., 1989). In this method, the *flp-1* gene was divided into two segments, each of which could be amplified by PCR with a pair of primers. For each segment, one primer annealed just beyond an end of the gene; the other primer was directed to the internal region of the gene where the mutation was to be introduced. The internal primers for the two segments carried the appropriately changed nucleotides. As a result, the two amplified segments overlapped slightly. When the two segments were added together, denatured, and reannealed, a single strand of one segment could anneal in the overlap region with the opposite strand of the other segment. One half of the hybrid molecules would have free 5′ ends facing the single-stranded portion. These are dead-ends. The other half would have free 3′ ends, which could prime DNA synthesis to give full-length duplex molecules with the mutation in both strands. By using the end-specific primers, the full-length, mutated *flp-1* gene could be amplified and cloned. The mutation was then confirmed in the clone by nucleotide sequencing.

2.2 Characterization of mutant Flp1 pilins

The mutant genes were inserted into a plasmid vector downstream of the IPTG (isopropyl β-D-thiogalactopyranoside)-inducible *tac* promoter. (The *tac* promoter is a strong promoter that is a hybrid of the *trp* and *lac* promoters. Like the *lac* promoter, the *tac* promoter is inhibited by the *Escherichia coli* LacI repressor protein, whose gene was already added to the plasmid cloning vector. LacI is inactivated by IPTG.) We wanted to know (1) if the mutant pilin had wild-type abundance, (2) if the mutant *flp-1* gene could complement a *flp-1* chromosomal mutant gene, (3) if the mutant pilin allowed Flp pili to be made, and (4) if any Flp pili assembled with the mutant pilin promoted adherence. Mutant Flp1 pilin abundance was indicated by immunological detection of pilin in protein extracts that were separated into bands by electrophoresis in a sodium dodecyl sulfate polyacrylamide gel (Western blot). Genetic complementation was detected by the conversion of the smooth-colony morphology of a *flp-1*- mutant strain to a rough-colony morphology reminiscent of the wild-type strain. The presence of Flp pili was shown by electron microscopy of *flp-1*- mutant cells carrying the mutant pilin gene. To quantify adherence, a slightly modified crystal violet assay of O'Toole and Kolter (1998) was used. Basically, wells of microtiter dishes were inoculated with the strains to be tested. Nonadherent cells were removed by washes. Adherent cells were then stained with crystal violet. After more washes to remove free crystal violet and any remaining nonadherent cells, the crystal violet in the adherent cells was then eluted with DMSO (dimethyl sulfoxide). The eluted crystal violet gave a color to the solution that was quantified with a spectrophotometer. Increasing crystal violet in the eluate is indicative of increasing adherence.

Each of our Flp1 mutants had one non-alanine residue changed to alanine. (See Table 1 for the single-letter and three-letter codes for the amino acids. Table 2 shows the residue change in the Flp1 pilin for each mutant and the phenotype.) In the mutants, every non-alanine residue of the mature Flp1 pilin was changed to alanine. The Flp1 prepilin has 75 residues; but, after cleavage, the mature Flp1 pilin has 49 residues. Nine residues are already alanine. The other 40 residues of the mature pilin were changed to alanine. The small size of mature Flp1 pilin made it reasonable to create this series of mutant pilins.

Amino acid	Single-letter code	Three-letter code
alanine	A	ala
arginine	R	arg
asparagine	N	asn
aspartic acid	D	asp
cysteine	C	cys
glutamic acid	E	glu
glutamine	Q	gln
glycine	G	gly
histidine	H	his
isoleucine	I	ile
leucine	L	leu
lysine	K	lys
methionine	M	met
phenylalanine	F	phe
proline	P	pro
serine	S	ser
threonine	T	thr
tryptophan	W	try
tyrosine	Y	tyr
valine	V	val

Table 1. Single-letter and three-letter codes for the amino acids.

Mutant Residue	Colony Morph.[a]	Adher.[b]	Auto-aggreg.[c]	Protein exp.[d]	Piliation[e]	Class[f]
V27A	S	-	-	+/-	+	IV
T28A	S	-	-	+	-	III
I30A	S	-	-	+	+	IV
E31A	S	-	-	+	-	III
Y32A	S	-	-	+/-	-	III
G33A	R	+	+	ND	ND	I

Mutant Residue	Colony Morph.[a]	Adher.[b]	Auto-aggreg.[c]	Protein exp.[d]	Piliation[e]	Class[f]
L34A	S	-	-	+/-	+	IV
I35A	S	-	-	-	-	II
I37A	S	-	-	+	-	III
V39A	R	+	+	ND	ND	I
V41A	R	+	+	ND	ND	I
L42A	R	+	+	ND	ND	I
I43A	S	-	-	+	+	IV
V44A	R	+	+	ND	ND	I
V46A	S	-	-	+	+	IV
F47A	S	-	-	+	-	III
Y48A	R	+	+	ND	ND	I
S49A	R	+	+	ND	ND	I
N50A	S	-	-	+	-	III
N51A	R	+	+	ND	ND	I
G52A	S	-	-	+	+	IV
F53A	R	+	+	ND	ND	I
I54A	S	-	-	-	-	II
N56A	R	+	+	ND	ND	I
L57A	S	-	-	+	+	IV
Q58A	R	+	+	ND	ND	I
S59A	R	+	+	ND	ND	I
K60A	R	+	+	ND	ND	I
F61A	S	-	-	+	-	III
N62A	R	+	+	ND	ND	I
S63A	R	+	+	ND	ND	I
L64A	S	-	-	+	-	III
S66A	R	+	+	ND	ND	I
T67A	R	+	+	ND	ND	I
V68A	R	+	+	ND	ND	I
S70A	R	+	+	ND	ND	I
N72A	R	+	+	ND	ND	I
V73A	S	-	-	+/-	-	III
T74A	R	+	+	ND	ND	I
K75A	S	-	-	+/-	+	IV

Table 2. Flp1 mutants and phenotypes.
[a] R, rough colony morphology; S, smooth colony morphology; [b] +, ≥60% wild-type adherence; -, <60% wild-type adherence; [c] +, autoaggregation; -, no autoaggregation; [d] +, wild-type Flp1 level; +/-, intermediate Flp1 level; -, no Flp1 observed; [e] +, pili observed; -, no pili observed; ND = not determined; [f] I-IV, phenotypic class

The mutant pilins were assayed for the phenotypes described above. We divided the mutant pilins into four phenotypic classes (Tables 2 and 3, Fig. 6). Class I Flp1 pilin mutants (21 in number) were indistinguishable from the wild-type pilin in our assays. In other *in vitro* assays or in animal experiments, some of these Class I mutants might show phenotypes that differ from those of the wild-type pilin. Six of the seven residues in the hydrophobic-region Class I mutants were originally glycine (1), valine (4), and leucine (1), all of which are similar to the alanine replacement. The seventh mutant in the hydrophobic region had alanine in place of tyrosine, a larger hydrophobic residue. There were two Class II mutants defined as showing no or very little pilin. Class III (9) and Class IV (8) mutants all showed abundant pilin, approximately equal to the abundance of the wild type pilin. However, electron micrographs showed that Class III pilins do not form pili. In contrast, Class IV mutant pilins could be assembled into pili; but the pili were nonadherent. Consequently, Class IV mutant cells did not autoaggregate, nor could a Class IV mutant strain form the tenacious Tad biofilm. One of the Class IV mutants (K75A) is particularly interesting because it produces curved and non-bundled pili.

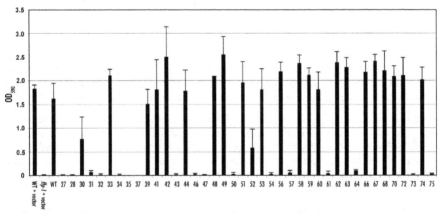

Fig. 6. Effect of mutant Flp1 pilins on adherence by *A. actinomycetemcomitans*.
Shown are the results of crystal violet assays for adherence. The first three results on the left are the controls: (1) positive controls, which are the wild-type *A. actinomycetemcomitans* strain (WT) with and without the empty plasmid used for cloning (vector) and (2) a negative control, which is a Flp1 pilin mutant strain (*flp-1*) with the empty vector. The mutant strain does not show tenacious adherence because it cannot make Flp pili. The numbers on the X-axis indicate the residues that were changed to alanine. The OD_{590} values on the Y-axis indicate relative amounts of crystal violet eluted from adherent cells (see text). The error bars show the calculated ranges of values from three experiments.

Because there is no 3D structure yet for Flp1 pilin, we used structure-prediction software with the sequence. Like other type IV pilins, the mature Flp1 pilin was predicted to be divided into a largely hydrophobic N-terminal domain and a distinct C-terminal domain. In Flp1, the C-terminal domain was predicted to contain an 11-residue amphipathic α-helix and a 12-residue, mostly-polar, C-terminal tail. It is thought that pilin subunits interact for polymerization in the hydrophobic region (Giltner et al., 2012). We therefore expected the "assembly" mutants (Class III) to be caused by alanine substitutions of residues predominantly in the hydrophobic N-terminal region. Conversely, the C-terminal region of a pilin is thought to interact with the environment (Giltner et al., 2012), so we expected Class IV mutants to occur mostly from alterations of C-terminal residues. However, we were somewhat surprised by the number of Class III mutants with a substitution in a C-terminal residue, and by the number of Class IV mutants with a change in an N-terminal residue.

Mutant Class	Adherence	Protein Expression	Piliation	Pilus Morphology
I	+	ND	ND	ND
II	-	-	ND	ND
III	-	+	-	NA
IV	-	+	+	Normal and altered

NA, "not applicable"; ND, "not determined."

Table 3. Summary of Flp1 mutant classes.

A previous study indicated that seven serine and asparagine residues in the C-terminal region are modified (Inoue et al., 2000). We found that only one substitution of those seven residues, a Class III mutant (N50A), gave a mutant phenotype. We do not know if the defect is due to the loss of modification at this residue or to the change in that amino acid.

Alanine-scanning mutagenesis has been an important step in beginning to understand the Flp1 pilin. We now have a collection of mutant pilins that can be studied for information on pilin stability, pili assembly, pili bundling, and pili adherence. These mutants will guide experiments in which substitutions can be made with amino acids that are of different sizes, have similar properties, or have very different properties. Similar experiments can also be done at the alanine residues.

A 3D structure of the Flp1 pilin is needed. There are a few structures of type IV pilins, most of which were determined from crystals formed after removing the N-terminal hydrophobic domain (Giltner et al., 2012). The Flp1 structure will clearly be different from the few structures of other type IV pilins. For example, Flp1 is 2 to 3-fold smaller than other type IV pilins. Also the mature Flp1 pilin has no cysteine residues, which are thought to be needed to form a disulfide bond to make the D region structure that seems to be conserved in type IV pilins. The phenotypes of the Flp1 mutants have also underscored the differences of Flp1 pilin and the "typical" type IV pilins. A structure would help us to understand (1) what is truly "typical" in type IV pilins, (2) the importance of certain Flp1 residues, and (3) the molecular basis of the phenotypes of the mutants.

3. TadZ

The *tadZ* gene has no known homolog, and it is unique to *tad* loci. The presence of a *tadZ* homolog is taken to indicate that a series of genes is, is part of, or once was a *tad* locus. In the *tad* locus of *A. actinomycetemcomitans*, *tadZ* is essential for the biogenesis of Flp pili and, therefore, for the tenacious biofilm. Our recent fluorescence-microscopy study of a TadZ-

EGFP (enhanced green fluorescent protein) fusion showed that TadZ protein localizes to the old cell pole (*i.e.*, opposite the pole formed by cell division) (Fig. 7) (Perez-Cheeks et al., 2012). It localized in the absence of any other protein encoded by the *tad* locus. (The TadZ-EGFP fusion was formed by eliminating the translational stop codon of the *tadZ* gene and attaching it in-frame to the coding sequence for EGFP minus its ribosome-binding-site and translational start codon.) In contrast, a TadA-EGFP fusion also localized to a pole, but only when TadZ protein was present. We proposed that TadZ is responsible for localizing the entire Tad secretory apparatus to a pole (Perez-Cheeks et al., 2012).

We did a large phylogenetic analysis and showed that the *tadZ* genes belong in the *parA/minD* superfamily of genes (Perez-Cheeks et al., 2012). The prototypical bacteriophage P1 ParA protein (and each of the various ParA-like proteins) is needed for the proper segregation of DNA at cell division. The *E. coli* MinD protein, which is thought to be the prototype for the other MinD-like proteins encoded by the family, has been studied extensively. It is needed to localize the septum properly at cell division. Other gene families in the *parA/minD* superfamily are the *nifH* family (named for nitrogen fixation), the *fleN* family (named for flagellar synthesis), and the *bcsQ* family (named for cellulose biosynthesis).

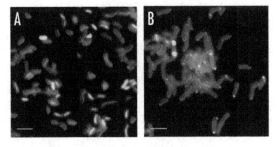

Fig. 7. Polar foci of the TadZ-EGFP fusion protein.
Shown are micrographs of *tadZ* mutant cells of *A. actinomycetemcomitans* slightly elongated by treatment with a sub-inhibitory concentration of the antibiotic piperacillin, stained with the membrane-specific fluorescent dye TMA-DPH [1-(4-trimethylammoniumphenyl)-6-phenyl-1,3,5-hexatriene p-toluenesulfonate] (fluoresces red), and containing a plasmid expressing (A) green fluorescent protein as a control or (B) a TadZ-EGFP fusion. The cells were illuminated with UV light. Polar foci are seen only in the cells expressing the TadZ-EGFP fusion protein (Perez-Cheeks et al., 2012).

3.1 The atypical Walker-like A box of TadZ proteins

Each of the protein products of the *parA/minD* superfamily of genes has a Walker-like A box [KGGXX(S/T)] (Fig. 8), which forms a structure involved in binding and hydrolyzing ATP. The Walker-like A box is a variation of the Walker A box [GXXGXGK(S/T)], which also allows a protein to bind and hydrolyze ATP. However, the ATPases of Walker A box proteins are considerably stronger than the ATPases of the Walker-like A box proteins. Each product of the *tadZ* gene family also has a Walker-like A box, but the TadZ Walker-like A box is unique. The second lysine (K6 in the numbering system of Fig. 8) is missing. We call the unique Walker-like A motif in TadZ proteins the "atypical Walker-like A box" (Perez-Cheeks et al., 2012).

	1	2	3	4	5	6	7
Walker-like	K	G	G	X	X	K	T/S
Ec MinD	K	G	G	V	G	K	T
P1 ParA	K	G	G	V	S	K	T
Aa TadZ	K	G	G	I	G	A	S
Cc TadZ	K	G	G	V	G	A	S

Fig. 8. Walker-like A boxes of proteins from the *parA/minD* superfamily.
Ec, *Escherichia coli*; P1, bacteriophage P1; Aa, *Aggregatibacter actinomycetemcomitans*;
Cc, *Caulobacter crescentus*

3.2 Phenotypes of mutants altered in the atypical Walker-like A box of TadZ

We wanted to know the effect of changing the residue in position 6 of the atypical Walker-like A box from alanine in the TadZ protein of *A. actinomycetemcomitans* (AaTadZ) to the lysine residue of the canonical Walker-like A box. We used overlap extension PCR (Section 2.1) to change the codon in *tadZ* for residue 155 of the AaTadZ protein. We cloned the mutant gene into a plasmid vector and expressed it from the *tac* promoter, as described in Section 2.2. The mutant and wild-type proteins were found to be equally abundant in cells. After we introduced the plasmid into a *tadZ* mutant strain, we assayed three phenotypes of *tad* locus function: (1) change in colony morphology of a *tadZ⁻* mutant from smooth (the mutant phenotype) to rough (the wild-type phenotype), (2) autoaggregation of cells, and (3) formation of the tenacious biofilm. Whereas wild-type AaTadZ protein was completely functional in all three phenotypes, the A155K mutant protein was completely defective in all

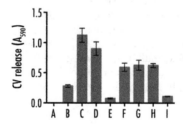

Fig. 9. Adherence of *A. actinomycetemcomitans* strains expressing the wild-type TadZ protein and mutants of TadZ.
Adherence was shown by the crystal violet (CV) assay described in the text. The Y-axis is the same as in Fig. 6. A-D are controls: (A) media alone; (B) the *tadZ⁻* mutant strain with an empty vector plasmid; (C) a wild-type strain with vector; and (D) the *tadZ⁻* mutant strain with a plasmid carrying *tadZ⁺*. E-I are *tadZ⁻* strains expressing mutant TadZ proteins: (E) TadZ A155K, (F) TadZ K150R, (G) TadZ K150A, (H) TadZ S156T, and (I) TadZ S156A. The error bars show the calculated ranges of values.

three phenotypes (see Fig. 9 for the adherence result). Therefore, the presence of alanine at position 6 of the Walker-like A box is essential for AaTadZ function.

We noticed that a common feature of the atypical Walker-like A boxes from other TadZ proteins was the absence of lysine at Walker-like box position 6, not the presence of alanine. The *tadZ* genes from various *tad* loci had codons for other amino acids at position 6.

We confirmed this property for AaTadZ. When we mutated the *tadZ* gene to substitute a glycine, valine, asaparagine, or serine residue for alanine in the protein, all the mutant proteins were functional.

Even though it seemed that the absence of lysine was the primary requirement for the position 6 residue of the Walker-like A box of TadZ proteins, other residues of the Walker-like A motif are conserved. This observation indicated that the other residues in the Walker-like A boxes of TadZ proteins are important for function. To test this, we made mutants of AaTadZ in which the lysine residue at Walker-like A box position 1 (AaTadZ K150) was changed to arginine or alanine. Likewise, we made mutants at position 7 (AaTadZ S156) with threonine or alanine in place of the conserved serine residue. The mutant AaTadZ proteins did not allow wild-type biofilm formation (Fig. 9). Strains with the K150R, K150A, and the S156T mutants showed some biofilm formation; but it was reduced relative to wild type. The S156A mutant was completely unable to adhere. We concluded that the other residues of the atypical Walker-like A box of TadZ proteins are important to a function leading to tenacious adherence and biofilm formation.

The MinD proteins from *E. coli* (Hu & Lutkenhaus, 1999; Raskin & de Boer, 1999) and *Neisseria gonorrhoea* (Ramirez-Arcos et al., 2002), the ParA protein from plasmid pB171 of *E. coli* (Ebersbach & Gerdes, 2002), and the ParA-like Soj protein of *Bacillus subtilis* (Marston & Errington, 1999; Quisel et al., 1999) form polar foci and oscillate from pole to pole in the cell. The mobility of MinD-like and ParA-like proteins depends on their ATPase activity (Lutkenhaus & Sundaramoorthy, 2003). Changing a residue in the Walker-like A box of ParA from pB171 (Ebersbach & Gerdes, 2001), of the MinD protein from *N. gonorrhoea* (Ramirez-Arcos et al., 2002), or of Soj (Quisel et al., 1999) causes the protein to lose its mobility. Our TadZ-EGFP experiments have led us to suggest that TadZ positions the Tad secretion apparatus to a pole. Our hypothesis requires that TadZ does not oscillate. Maybe AaTadZ is a natural variant selected to allow it to remain stationary at a pole. The rest of the Walker-like A box may be needed to bind ATP, which was found in crystals of TadZ from *Eubacterium rectale* (Xu et al., 2012). The binding of ATP is not for the polar localization nor for dimer formation (see Sections 3.3 and 3.4 below). We note that ATP binding is necessary for the interaction of MinD with MinC and MinE (Hayashi et al., 2001; Hu et al., 2002), of Soj with SpoOJ (Marston & Errington, 1999; Quisel et al., 1999), and of bacteriophage P1 ParA with ParB (Bouet & Funnell, 1999). Perhaps ATP is needed for TadZ to interact with one or more proteins of the Tad secretion apparatus.

3.3 Mutants altered in the atypical Walker-like A box of TadZ localize properly

We used PCR to amplify the *tadZ* genes that encoded the atypical Walker-like A box mutant proteins. The amplified mutant genes were used to make mutant TadZ-EGFP fusions, as described for the wild-type gene in Section 3. All of the mutant fusion proteins formed normal-appearing fluorescent foci at the old poles of cells. The absence of a defect in focus formation or location was particularly striking for the A155K and S156A mutants. A mutant TadZ protein with either allele was completely defective in the phenotypes we assayed (Fig. 9 and Section 3.2). In fact, the mutant fusions consistently showed a higher number of cells with polar foci than was seen with the wild-type TadZ-EGFP fusion. We do not understand the significance of the increase, but we concluded that the residues of the atypical Walker-like A box are not needed for the polar localization of TadZ.

3.4 Mutants altered in the atypical Walker-like A box of TadZ form dimers

We used a bacterial reporter strain to indicate TadZ dimer formation *in vivo* (Hu et al., 2000). The basis of the reporter is the following. Bacteriophage λ cI protein represses the λ p_R promoter. In the reporter strain, the E. *coli lacZ* gene, which encodes β-galactosidase, was fused to p_R. Because there is a convenient colorimetric assay for β-galactosidase, expression of *lacZ* from the synthetic p_R-*lacZ* operon can be measured. The level of β-galactosidase is a function of p_R activity and is, therefore, an indication of cI activity. cI protein is a dimer (Pabo et al., 1979). Each monomeric polypeptide has an N-terminal domain for DNA binding and a C-terminal domain for dimerization. If the C-terminal domain is removed, the N-terminal DNA-binding domain (λ cIDB) is unable to function because it cannot form dimers. Fusing a polypeptide that can dimerize to λ cIDB can make a functional repressor and reduce expression of p_R-*lacZ*.

We created a chimeric gene that encoded a fusion of the coding regions for λ cIDB and TadZ. The product of the fusion repressed p_R-*lacZ* and indicated that TadZ can form dimers (Table 4) (Perez-Cheeks et al., 2012). Knowing that TadZ can dimerize, we then asked if the atypical Walker-like A box mutants are proficient or defective in the dimerization activity (Perez-Cheeks et al., 2012). To do this, it was necessary to make chimeras of the coding region for λ cIDB and the mutant *tadZ* genes. Each fusion protein was then tested in the reporter strain for the ability to repress p_R-*lacZ*. Each mutant fusion protein made a functional repressor that was as good as the fusion with wild-type TadZ (Table 4). We concluded that individual residues of the atypical Walker-like A box of TadZ are not needed for dimer formation.

Protein	% repression of p_R-*lacZ*
Empty vector	0.0 ± 0.0
λcIDB	9.7 ± 4.0
λcI	66.0 ± 1.4
λcIDB-TadZ	60.1 ± 0.7
λcIDB-TadZ K150R	49.0 ± 1.9
λcIDB-TadZ K150A	60.3 ± 1.5
λcIDB-TadZ A155K	49.7 ± 2.1
λcIDB-TadZ S156T	53.7 ± 0.8
λcIDB-TadZ S156A	57.6 ± 4.2

Table 4. *In vivo* dimerization assay results for wild-type TadZ and mutant TadZ proteins.

4. Regulation of the *tad* locus

Logic and evidence indicate that the expression of the *tad* locus genes in *A. actinomycetemcomitans* is controlled. The Tad⁻ variants that spontaneously arise from Tad⁺ parents form larger colonies on solid medium, grow to a higher density in liquid medium, and have a shorter generation time than do their parents. We conclude that it is energetically expensive for a cell to make Flp pili. Therefore, it would be advantageous to the cell for *tad* locus expression to be regulated. One way to accomplish this is to control *tad* gene transcription.

There is evidence for transcriptional regulation of *tad* loci. *P. aeruginosa* has a *tad* locus, and it is regulated by the PprA-PprB two-component system (Bernard et al., 2009). The *pprA* and *pprB* genes map within a locus that has five contiguous, but divergent, transcriptional units. Four encode the *tad* genes and *pprA*; the fifth encodes *pprB* only. After being activated by the histidine kinase (PprA), the response regulator (PprB) binds to the promoters and activates transcription of the *tad* genes. In another example, the expression of the *tad* genes (the *flp* operon) in the human pathogen *Haemophilus ducreyi* is affected by the CpxR-CpxA two-component system (Labandeira-Rey et al., 2010). Overexpression or constitutive activation of the response regulator CpxR causes repression of *flp* operon transcription and a reduction in the level of Flp1 protein. The authors suggested that activated CpxR directly represses transcription of the *flp* operon because CpxR bound to a target in the *flp* promoter region.

For *A. actinomycetemcomitans*, Kaplan et al. (2003) have shown the presence of probable nonadherent cells within a Tad⁺ colony. They found sequestered, loosely packed, non-aggregating, and probably nonadherent cells in an adherent colony of Tad⁺ cells. The authors suggested that transiently nonadherent cells are produced by the colony as part of a developmental pathway to expand the biofilm. Temporary nonadherence would allow cells to escape and seed new adherent colonies that are needed for the biofilm to spread. One possibility is that the nonadherent and non-autoaggregating cells lack Flp pili due to the inhibition of *tad* gene transcription. Indeed, there is a provocative 31-bp (base-pair) inverted repeat (IR) adjacent to the *tad* promoter (Fig. 3B) (G. Hovel-Miner, P. Planet, and D. Figurski, unpublished results). (The 31-bp IR has a spacer of 11 bp flanked by 10-bp arms that are inverted complementary repeats of each other.) The IR is conserved in all six serotypes of *A. actinomycetemcomitans*, hinting that it is important to this organism. However, the function of the IR is currently unknown. Because the *tad* promoter is very strong (Kram et al., 2008), if the IR is a binding site for a protein that regulates transcription, it seems likely that the protein that binds IR would be a repressor that reduces *tad* transcription.

Our studies have indicated that *tad* locus transcription is regulated by a termination cascade to maintain the correct stoichiometry of the *tad* gene products (Kram et al., 2008). We isolated three transcriptional terminators (T). T1 and T3 are factor-independent terminators, whereas T2 is a Rho-dependent terminator. RNA sizes and results from a *lacZ* transcriptional reporter indicated that T1 accounts for ~99% termination and is located after the pilin gene, *flp-1*, and before *flp-2*. T2 is located between *tadV* and *rcpC* and seems to terminate ~36% of the remaining transcripts in our assay. T3 terminates transcription after *tadG*, i.e., at the end of the locus opposite the promoter. Rough RNA quantitation indicates that for every full-length *tad* transcript, there are ~1.5 *flp-1/flp-2/tadV* transcripts and ~160 *flp-1* transcripts. This means that, assuming all *tad* mRNAs are translated with equal efficiencies, for every Tad secretion apparatus, there are about 50% more TadV protease molecules to process the abundance of Flp1 prepilin protein.

4.1 Tad⁺ and Tad⁻ bacteria make different biofilms

Biofilm formation depends on adherence. Biofilms with different characteristics may indicate different modes of adherence. Tad⁺ *A. actinomycetemcomitans* synthesize a pili-based, resilient biofilm. Tad⁻ variants have been reported to form weak biofilms (Inoue et al., 2003). We noticed that cells of Tad⁻ variants displayed adherence when the culture was handled gently (Fig. 10). Most or all of the biofilm of a Tad⁻ strain is lost under the conditions we use to allow the biofilm of a Tad⁺ strain to remain intact. Three-dimensional light microscopy showed that the biofilm of a Tad⁻ strain is very different from the biofilm of its Tad⁺ parent. Biofilms of Tad⁺ strains showed distinct, tightly packed microcolonies of cells. In contrast, the biofilms of Tad⁻ strains showed loosely packed cells in an extracellular matrix that stained readily with DAPI (4',6-diamidino-2-phenylindole, a fluorescent DNA stain). The biofilm of the Tad⁻ strain showed cells in structures that were interpreted to be columns and mushroom shapes. We suggested that the Tad⁻ biofilm resembles a "typical" biofilm.

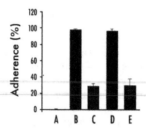

Fig. 10. Adherence of isogenic Tad⁺ and Tad⁻ strains of *A. actinomycetemcomitans*.
Shown are results from the crystal violet assay for adherence (see text) for two isogenic pairs of *A. actinomycetemcomitans* strains (B, C and D, E). A is a no-cells control. B and D are the Tad⁺ parents: DF2200N (serotype a) and CU1000N (serotype f). The two spontaneous Tad⁻ (smooth-colony) variants are DF2261N (C) and CU1060N (E). The adherence of DF2200N is taken as 100%. The error bars show the Standard Deviation calculated from the results of three experiments. Some adherence of the variants can be detected in conditions that favor the tenacious adherence of Tad⁺ strains.

4.2 Choosing proteins that may be needed for adherence to inert surfaces

We sought to find a non-*tad*-locus protein needed for adherence in *A. actinomycetemcomitans*. We identified what we thought were prime candidates for adherence-essential proteins in *A. actinomycetemcomitans* (Table 5). We considered two-component systems because the two known examples of *tad* regulation by a non-*tad* determinant involve two-component systems. Only four two-component systems are predicted to be encoded by the genome of *A. actinomycetemcomitans*: ArcAB, CpxRA, NarPQ, and QseBC. We also selected two other proteins that we thought might be required for adherence: OxyR and PgaC. OxyR was chosen because *Aggregatibacter aphrophilus*, a bacterium closely related to *A.*

actinomycetemcomitans, has a nearly identical *tad* locus (P. Planet, C. Sheth, and D. Figurski, unpublished results; Di Bonaventura et al., 2009). *A. aphrophilus* makes more Flp pili in higher oxygen than in lower (S. Kachlany, C. Sheth, and D. Figurski, unpublished results). Therefore, we selected OxyR, a transcriptional activator that responds to oxygen (Bauer et al., 1999), because the *tad* genes in this bacterium appear to respond to the presence of oxygen. In addition, there is evidence that OxyR induces the adhesin *apiA* in *A. actinomycetemcomitans* (Ramsey & Whiteley, 2009). PGA (poly-β-1,6-N-acetyl-D-glucosamine) is an extracellular polymer that is synthesized by *A. actinomycetemcomitans* and is associated with the biofilm of Tad⁺ cells (Izano et al., 2008). We considered PgaC, an enzyme in the pathway for PGA synthesis, to be a candidate for a protein that affects a biofilm in *A. actinomycetemcomitans*.

Gene	Function
arcB	Two-component sensor kinase; response to anaerobic and aerobic stress
cpxR	Two-component response regulator; response to outer membrane stress
narP	Two-component response regulator; response to nitrate and nitrite stress
qseB	Two-component response regulator; response to quorum sensing
oxyR	Transcriptional regulator; response to redox stress
pgaC	Integral membrane glycosyltransferase; required for PGA biosynthesis

Table 5. Selected putative regulators of adherence by *A. actinomycetemcomitans*.

4.3 Mutagenesis of possible adherence-required genes by allelic exchange

We constructed a series of six mutant Tad⁺ strains and six mutant Tad⁻ strains. Each strain had a mutation in one of our six candidates genes (Section 4.2). We asked if any of the mutants was defective in the strain's biofilm. The mutant strains were made by allelic exchange, *i.e.*, the mutant gene was substituted for the wild-type gene in the chromosome (Fig. 11). In allelic exchange, the mutant gene is marked (usually with an antibiotic resistance gene) and introduced into a wild-type strain. Homologous recombination allows the mutant gene and its marker to replace the wild-type gene in the chromosome. If the gene is not essential for viability, strains in which the exchange has taken place can be selected because they are able to grow in the presence of the marker antibiotic. Strains that have not done the exchange will not grow in the presence of the antibiotic because the introduced DNA cannot replicate.

We used plasmid pMB78 (Bhattacharjee et al., 2007), a "suicide" plasmid, *i.e.*, one that can replicate or be maintained in one host and not in another. Plasmid pMB78 can be maintained in *E. coli*, but not in *A. actinomycetemcomitans*. The wild-type gene was amplified from the chromosome of *A. actinomycetemcomitans* and cloned into pMB78 in *E. coli*. To construct the mutant allele, the internal part of the wild-type gene was deleted and replaced with the gene for kanamycin resistance. Plasmid pMB78 has the uptake signal sequence (USS) needed for transformation of *A. actinomycetemcomitans*. After transformation and recombination, the recombinants (*i.e.*, the mutant strains) were selected by their growth on medium containing kanamycin. The presence of the mutant alleles was confirmed by PCR and gel electrophoresis (Fig. 12A).

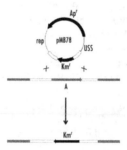

Fig. 11. Schematic for allelic exchange.

pMB78 is an ampicillin resistance (Ap^r)-encoding suicide plasmid that cannot be maintained in *A. actinomcetemcomitans* (see text) (Bhattacharjee et al., 2007). The wild-type gene is cloned and mutated by insertion of DNA encoding (in this example) kanamycin resistance (Km^r). After recombination ("X") in the homologous regions (white boxes), the mutated gene replaces the wild-type gene in the chromosome. The mutant strains (recombinants) can grow in the presence of kanamycin. "rep" is "replication region"; "USS," "uptake signal sequence," which is needed for transformation of *A. actinomycetemcomitans* (Thomson et al., 1999).

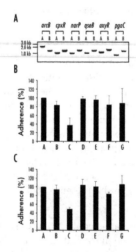

Fig. 12. Directed mutations and their effects on the adherence of Tad$^+$ and Tad$^-$ strains of *A. actinomycetemcomitans*.

Panel A shows agarose gel electrophoresis of PCR products of the wild-type gene and the mutant gene after recombination for each of the six possible regulators of adherence. Panels B and C show the results of the crystal violet assay for adherence (see text) of the isogenic Tad$^+$ and Tad$^-$ strains, respectively. In Panels B and C, adherence of the parent strain (A) is normalized to 100%. The mutant strains are defective in *arcB* (B), *cpxR* (C), *narP* (D), *qseB* (E), *oxyR* (F), and *pgaC* (G).

Adherence of the Tad⁺ (Fig. 12B) and Tad⁻ (Fig. 12C) mutants was quantified by the crystal violet assay (Section 2.2). One mutation, Δ*cpxR*::Kmʳ ("::" indicates "insertion of"), consistently caused a significant decrease in adherence (relative to the wild-type strain) for both the Tad⁺ and Tad⁻ strains (Fig. 12 B and C, respectively). This result showed that the Δ*cpxR*::Kmʳ mutation affected adherence in both types of strains. An important confirmation was to ask if adding the intact *cpxR*⁺ gene *in trans* restored adherence to the mutant strains (genetic complementation). If it did, then we could conclude that the defect in the mutant strain was the result of not having a functional CpxR protein. The *cpxR*⁺ gene was added *in trans* as a cloned gene on a plasmid that replicates in *A. actinomycetemcomitans*. We were surprised to learn that *cpxR*⁺ *in trans* restored adherence to the Tad⁻ strain (Fig. 13A), but not to the Tad⁺ strain (Fig. 13B). For the Tad⁺ strain, genetic complementation occurred only when the complete *cpxRA* operon was added *in trans*, indicating the need for CpxA in Tad⁺ adherence (Fig. 13B). More needs to be done to make this a solid conclusion. However, we can conclude that both types of adherence in *A. actinomycetemcomitans* respond to a function encoded by the *cpxRA* operon. We wish to point out that adherence in *A. actinomycetemcomitans* is positively regulated by the *cpxRA* operon, and we do not know whether it functions directly or indirectly to regulate adherence. In contrast, the activation of CpxR in *H. ducreyi* leads the molecule to bind to the *tad* (*flp* operon) promoter as a negative regulator (Labandeira-Rey et al., 2010).

5. A strategy for making precise genomic deletions: the new Vector Excision (VEX) method

In Section 4.3, allelic exchange was described as a method of exchanging a wild-type segment of the chromosome with a mutated segment. In the examples given, a single gene was mutated. Allelic exchange can also be used for more than one gene. However, because the technique usually relies on standard molecular cloning methods, allelic exchange often becomes more difficult with larger segments. (See the chapter by Gerlach et al. for a discussion of recombineering – a method of cloning that overcomes the limitation caused by the locations of restriction endonuclease cleavage sites.) Another problem can be caused by the insertion of the marker, which can affect the expression of downstream genes. A method based on site-specific recombination has been developed to remove the marker (Datsenko and Wanner, 2000).

We have developed a straightforward method for making chromosomal deletions that can be any size [one bp to several kb (kilobase pairs)] and do not affect the expression of downstream genes. The strategy is based on the Vector Excision (VEX) method that we developed (Ayres et al., 1993). With VEX, the deleted portion can also be "captured" on a self-transmissible, broad-host-range plasmid. After transfer to another bacterium, the expression and/or functions of the captured genes can be assessed in a different host (see the chapter by Wilson et al.).

The new VEX strategy is illustrated here as a deletion (~11 kb) of all the *tad* genes of *A. actinomycetemcomitans* (Fig. 14). A "double cointegrate" is generated by two cycles of homologous recombination. Essentially the double cointegrate is formed by two successive allelic exchanges (Section 4.3). The plasmid suicide vector we used was based on pBBR1MCS (Kovach et al., 1994), which replicates in *E. coli*, but not in *A.*

actinomycetemcomitans. The plasmid can be mobilized into both species by conjugation, and the USS required for transformation of *A. actinomycetemcomitans* was added along with the gene for ampicillin resistance by recombineering. Thus, the vector can be introduced into *A. actinomycetemcomitans* either by transformation (as was done here) or by conjugative mobilization (crucial for strains that transform poorly or not at all).

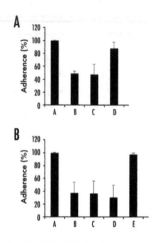

Fig. 13. Genetic complementation of the adherence defect from the *cpxR*- mutation in isogenic Tad- and Tad+ strains of *A. actinomycetemcomitans*.
Shown are crystal violet assays for adherence (see text). Panel A is the Tad- strain; panel B is the Tad+ strain. Column A in panel A is the smooth-colony variant with the wild-type *cpxR* gene. Its adherence is normalized to 100%. Column B is the *cpxR*- mutant strain; column C, the *cpxR*- mutant strain with the empty plasmid vector; and column D, the *cpxR*- mutant strain with the *cpxR*+ plasmid. Column A in panel B is the Tad+ parent. B-D are analogous to those in Panel A, but with the Tad+ stain. Column E is the *cpxR*- mutant strain with a *cpxRA*+ plasmid. Note that complementation of the *cpxR*- mutation in the Tad+ strain requires *cpxRA*+ *in trans*, whereas *cpxR*+ *in trans* is sufficient for complementation of the *cpxR*- mutation in the Tad- strain.

The purpose of the two homologous recombination events is to integrate two directly repeated *loxP* sites. The homology cassette determines where the *loxP* sequence is inserted into the chromosome. In our experiment, one *loxP* sequence was inserted in the *flp-1* gene; and the other, in *tadG*. The "left" (in Fig. 14) *loxP* sequence was marked by the *aacC1* gene for gentamicin resistance. The "right" *loxP* sequence was marked by *aadA*, the gene for spectinomycin resistance. Each *loxP* cassette was formed by cloning three fragments into the multiple cloning site (MCS) of the vector: (1) homology region I (HRI), (2) *loxP* and *aacC1* or *aadA*, and (3) homology region II (HRII). The fragments were generated by PCR. The 34-bp *loxP* sequence was added to the appropriate primer for fragment (2). Fragments were

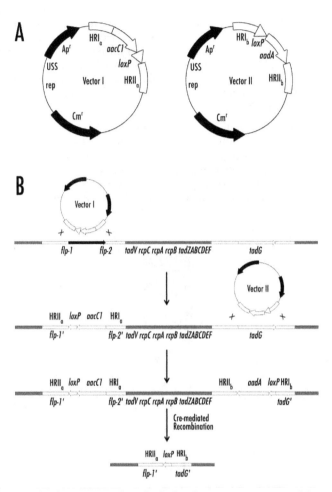

Fig. 14. Schematic for the "new VEX" method of making precise genomic deletions.
Panel A shows the two plasmids that will be used in the construction of the double
cointegrate. Vectors I and II are suicide plasmids that can replicate in *E. coli*, but not in *A.
actinomycetemcomitans*. They are made from the same plasmid. The symbols and
abbreviations are the following: the *loxP* cassettes (with cloned fragments) are in white;
"Ap^r," ampicillin resistance; "Cm^r," chloramphenicol resistance; "rep," replicon; "USS,"
uptake signal sequence for transformation of *A. actinomycetemcomitans*; "HR," homology
region; "*aacC1*," the gene for gentamicin resistance; "*aadA*," the gene for spectinomycin
resistance; and "*loxP*," a 34-bp sequence that is the target for Cre-mediated site-specific
recombination. Panel B depicts the two homologous recombination events and resolution by
the Cre recombinase to delete the genes of the *tad* locus in *A. actinomycetemcomitans*. See the
text for details.

cloned with the following restriction endonucleases: *XbaI* and *Bam*HI for fragment (1), *Bam*HI and *SalI* for fragment (2), and *SalI* and *KpnI* for fragment (3). The site-specific recombinase Cre binds to a *loxP* sequence (Sternberg and Hoess, 1983), forms a synapse with another Cre-bound *loxP* sequence, and catalyzes recombination between *loxP* sequences. Cre-mediated recombination is very efficient (>95%). Directly repeated *loxP* sites cause the intervening DNA to cyclize and be deleted by the action of Cre. Inverted repeats of *loxP* sites lead to inversion of the intervening DNA. Cre was supplied to the double cointegrate by conjugation with a *cre*-encoding plasmid.

After the Cre-mediated deletion (resolution) (Fig. 15), a single *loxP* sequence is left in place of the deleted DNA (Fig. 14). In the *flp-1* to *tadG* deletion, ~11 kb of DNA was removed, whereas one 34-bp *loxP* sequence was inserted. Our results show that a double *loxP*-containing homology cassette with antibiotic resistance genes blocks transcription of a downstream gene. In contrast, the single *loxP* sequence present after resolution allows transcription and expression of the downstream gene (T. McConville and D. Figurski, unpublished results).

1 2 3

Fig. 15. Deletion of the *tad* locus of *A. actinomycetemcomitans*.
Shown is an agarose gel of PCR products that show the wild-type *tad* locus (1) and a genomic deletion of ~11 kb that removes all the genes of the *tad* locus (3). The deletion was made using the new VEX strategy shown in Fig. 14. Lane 2 shows DNA markers of different molecular weights.

6. Conclusion

The 14-gene *tad* locus of *A. actinomycetemcomitans* is needed for the synthesis and secretion of Flp pili. The ability of *A. actinomycetemcomitans* to cause periodontal disease depends on Flp pili-mediated tenacious adherence in the oral cavity. (Flp pili likely mediate colonization of teeth.) We think we may understand the function of Flp pili in the etiology of oral disease caused by *A. actinomycetemcomitans*. However, *tad* loci occur in nearly 40% of prokaryotes. We have suggested that *tad* loci help prokaryotes colonize a niche (Kachlany et al., 2001a). For *A. actinomycetemcomitans*, a major niche is known to be the oral cavity. Tenacious adherence may be a property of *A. actinomycetemcomitans* because it must be able to colonize in the presence of extensive normal flora. Therefore, the strong phenotype of tenacious adherence may be a special property of the *tad* locus of *A. actinomycetemcomitans*. The strong phenotype allowed us to discover the *tad* genes. The phenotypes may be important, but more subtle, in other organisms.

Several proteins are unique products of *tad* loci. Such proteins may be good targets for therapeutic drugs. For example, *tadZ* is present in every *tad* locus. Our genetic studies in *A. actinomycetemcomitans* have shown that the *tadZ* gene is required for the function of its *tad* locus. We believe it is likely to be important in all *tad* loci. Therefore, a drug specific for TadZ protein might inactivate all *tad* loci. We have shown that the *tad* locus of *A. actinomycetemcomitans* is required for colonization of the mouth and periodontal disease. If we are correct that *tad* loci in other prokaryotes are needed to colonize their specific niches, inhibiting TadZ to inactivate *tad* loci may be useful in preventing and/or minimizing some diseases. Such a drug that targets a non-essential colonization factor should also be largely refractory to the selective pressure for the emergence of resistant strains.

Genetic studies of *tad* loci and the development of new genetic tools are helping us to determine and understand (1) the functions of the individual *tad* genes, (2) whether and how the various proteins encoded by a *tad* locus act to colonize a specific niche, and (3) the importance of the *tad* genes for *A. actinomycetemcomitans* and for other prokaryotes.

7. Acknowledgments

We are grateful to David Furgang, Scott Kachlany, Jeff Kaplan, Mari Karched, Paul Planet, Helen Schreiner, Kabilan Velliyagounder, James Wilson, and Gang Yue for discussions and/or technical help. We appreciate the efforts of Oliver Jovanovic on the figures. The work was funded by grants from the National Institutes of Health (NIH), USA. Additional NIH funding supported B.A.P., V.W.G., and K.E.K.

8. References

Ayres, E.K., Thomson, V.J., Merino, G., Balderes, D. & Figurski, D.H. (1993). Precise deletions in large bacterial genomes by vector-mediated excision (VEX). The *trfA* gene of promiscuous plasmid RK2 is essential for replication in several gram-negative hosts, *J. Mol. Biol.* 230, 174-185.

Bauer, C.E., Elsen, S. & Bird, T.H. (1999). Mechanisms for redox control of gene expression, *Annu. Rev. Microbiol.* 53, 495-523.

Bernard, C.S., Bordi, C., Termine, E., Filloux, A. & de Bentzmann, S. (2009). Organization and PprB-dependent control of the *Pseudomonas aeruginosa tad* Locus, involved in Flp pilus biology, *J. Bacteriol.* 191, 1961-1973.

Bhattacharjee, M.K., Kachlany, S.C., Fine, D.H. & Figurski, D.H. (2001). Nonspecific adherence and fibril biogenesis by *Actinobacillus actinomycetemcomitans*: TadA protein is an ATPase, *J. Bacteriol*. 183, 5927-5936.

Bhattacharjee, M.K., Fine, D.H. & Figurski, D.H. (2007). *tfoX (sxy)*-dependent transformation of *Aggregatibacter (Actinobacillus) actinomycetemcomitans*, *Gene* 399, 53-64.

Bouet, J.Y. & Funnell, B.E. (1999). P1 ParA interacts with the P1 partition complex at parS and an ATP-ADP switch controls ParA activities, *EMBO J*. 18, 1415-1424.

Clock, S.A., Planet, P.J., Perez, B.A. & Figurski, D.H. (2008). Outer membrane components of the Tad (tight adherence) secreton of *Aggregatibacter actinomycetemcomitans*, *J. Bacteriol*. 190, 980-990.

Datsenko, K.A. & Wanner, B.L. (2000). One-step inactivation of chromosomal genes in *Escherichia coli* K-12 using PCR products, *Proc. Natl. Acad. Sci. USA* 97(12):6640-5.

de Bentzmann, S., Aurouze, M., Ball, G. & Filloux, A. (2006). FppA, a novel *Pseudomonas aeruginosa* prepilin peptidase involved in assembly of type IVb pili, *J. Bacteriol*. 188, 4851-4860.

De Rycke, J. & Oswald, E. (2001). Cytolethal distending toxin (CDT): a bacterial weapon to control host cell proliferation? *FEMS Microbiol. Lett*. 203, 141–148.

Di Bonaventura, M.P., DeSalle, R., Pop, M., Nagarajan, N., Figurski, D.H., Fine, D.H., Kaplan, J.B. & Planet, P.J. (2009). Complete genome sequence of *Aggregatibacter (Haemophilus) aphrophilus* NJ8700, *J. Bacteriol*. 191, 4693-4694.

DiFranco, K.M., Gupta, A., Galusha, L.E., Perez, J., Nguyen, T.V., Fineza, C.D. & Kachlany, S.C. (2012). Leukotoxin (Leukothera®) targets active leukocyte function antigen-1 (LFA-1) protein and triggers a lysosomal mediated cell death pathway, *J. Biol. Chem*. 287, 17618-17627.

Ebersbach, G. & Gerdes, K. (2001). The double *par* locus of virulence factor pB171: DNA segregation is correlated with oscillation of ParA, *Proc. Natl. Acad. Sci. USA* 98, 15078–15083

Fine, D.H., Furgang, D., Kaplan, J., Charlesworth, J. & Figurski. D.H. (1999a). Tenacious adhesion of *Actinobacillus actinomycetemcomitans* strain CU1000 to salivary-coated hydroxyapatite. *Arch. Oral. Biol*. 44, 1063-1076.

Fine, D.H., Furgang, D., Schreiner, H.C., Goncharoff, P., Charlesworth, J., Ghazwan, G., Fitzgerald-Bocarsly, P. & Figurski, D.H. (1999b). Phenotypic variation in *Actinobacillus actinomycetemcomitans* during laboratory growth: implications for virulence, *Microbiology* 145, 1335-1347.

Fine, D.H., Goncharoff, P., Schreiner, H., Chang, K.M., Furgang, D. & Figurski, D. (2001). Colonization and persistence of rough and smooth colony variants of *Actinobacillus actinomycetemcomitans* in the mouths of rats, *Arch. Oral Biol*. 46, 1065-1078.

Fine, D.H., Velliyagounder, K., Furgang, D. & Kaplan, J.B. (2005) The *Actinobacillus actinomycetemcomitans* autotransporter adhesin Aae exhibits specificity for buccal epithelial cells from humans and old world primates, *Infect. Immun*. 73, 1947-1953.

Fine, D.H., Kaplan, J.B., Kachlany, S.C. & Schreiner, H.C. (2006). How we got attached to *Actinobacillus actinomycetemcomitans*: a model for infectious diseases, *Periodont. 2000* 42, 114–157.

Fine, D.H., Kaplan, J.B., Furgang, D., Karched, M., Velliyagounder, K. & Yue, G. (2010). Mapping the epithelial-cell-binding domain of the *Aggregatibacter actinomycetemcomitans* autotransporter adhesin Aae, *Microbiology* 156, 3412-3420.

Fives-Taylor, P.M., Meyer, D.H., Mintz, K.P. & Brissette, C. (1999). Virulence factors of *Actinobacillus actinomycetemcomitans*, *Periodontol. 2000* 20, 136-167.

Fong, K.P., Pacheco, C.M., Otis, L.L., Baranwal, S., Kieba, I.R., Harrison, G., Hersh, E.V., Boesze-Battaglia, K. & Lally, E.T. (2006). *Actinobacillus actinomycetemcomitans* leukotoxin requires lipid microdomains for target cell cytotoxicity, *Cell Microbiol.* 8, 1753-1767.

Fuller, T.E., Kennedy, M.J. & Lowery, D.E. (2000). Identification of *Pasteurella multocida* virulence genes in a septicemic mouse model using signature-tagged mutagenesis, *Microb. Pathog.* 29, 25-38.

Genco, R.J., Christersson, L.A. & Zambon, J.J. (1986). Juvenile periodontitis, *Int. Dent. J.* 36, 168-176.

Giltner, C.L., Nguyen, Y. & Burrows, L.L. (2012). Type IV Pilin Proteins: Versatile Molecular Modules, *Microbiol. Mol. Biol. Rev.* 76, 740-772.

Haase, E.M., Zmuda, J.L. & Scannapieco, F.A. (1999). Identification and molecular analysis of rough-colony-specific outer membrane proteins of *Actinobacillus actinomycetemcomitans*, *Infect. Immun.* 67, 2901-2908

Haase, E.M., Stream, J.O. & Scannapieco, F.A. (2003). Transcriptional analysis of the 5' terminus of the *flp* fimbrial gene cluster from *Actinobacillus actinomycetemcomitans*, *Microbiology* 149(Pt 1), 205-215.

Hayashi, I., Oyama, T. & Morikawa, K. (2001). Structural and functional studies of MinD ATPase: implications for the molecular recognition of the bacterial cell division apparatus, *EMBO J.* 20, 1819-1828.

Henderson, B., Wilson, M., Sharp, L. & Ward, J.M. (2002). *Actinobacillus actinomycetemcomitans*, *J. Med. Microbiol.* 51, 1013-1020.

Henderson, B., Nair, S.P., Ward, J.M. & Wilson, M. (2003). Molecular pathogenicity of the oral opportunistic pathogen *Actinobacillus actinomycetemcomitans*, *Annu. Rev. Microbiol.* 57, 29-55.

Henderson B., Ward J.M. & Ready D. (2010). *Aggregatibacter* (*Actinobacillus*) *actinomycetemcomitans*: a triple A* periodontopathogen? *Periodontol. 2000* 54, 78-105.

Ho, S.N., Hunt, H.D., Horton, R.M., Pullen, J.K. & Pease, L.R. (1989). Site-directed mutagenesis by overlap extension using the polymerase chain reaction, *Gene* 77, 51-59.

Hu, J.C., Kornacker, M.G. & Hochschild, A. (2000). *Escherichia coli* one- and two-hybrid systems for the analysis and identification of protein-protein interactions, *Methods* 20, 80-94.

Hu, Z. & Lutkenhaus, J. (1999). Topological regulation of cell division in *Escherichia coli* involves rapid pole to pole oscillation of the division inhibitor MinC under the control of MinD and MinE, *Mol. Microbiol.* 34, 82–90.

Hu, Z., Gogol, E.P. & Lutkenhaus, J. (2002). Dynamic assembly of MinD on phospholipid vesicles regulated by ATP and MinE, *Proc. Natl. Acad. Sci. USA* 99, 6761-6766.

Inoue, T., Tanimoto, I., Ohta, H., Kato, K., Murayama, Y. & Fukui, K. (1998). Molecular characterization of low-molecular-weight component protein, Flp, in *Actinobacillus actinomycetemcomitans* fimbriae, *Microbiol. Immunol.* 42, 253–258.

Inoue, T., Ohta, H., Tanimoto, I., Shingaki, R. & Fukui, K. (2000). Heterogeneous post-translational modification of *Actinobacillus actinomycetemcomitans* fimbrillin, *Microbiol. Immunol.* 44, 715-718.

Izano, E.A., Sadovskaya, I., Wang, H., Vinogradov, E., Ragunath, C., Ramasubbu, N., Jabbouri, S., Perry, M.B. & Kaplan, J.B. (2008). Poly-N-acetylglucosamine mediates biofilm formation and detergent resistance in *Aggregatibacter actinomycetemcomitans*, *Microb. Pathog.* 44, 52-60.

Jiang, X., Ruiz, T. & Mintz, K.P. (2012). Characterization of the secretion pathway of the collagen adhesin EmaA of *Aggregatibacter actinomycetemcomitans*, *Mol Oral Microbiol.* 27, 382-396.

Kachlany, S.C., Planet, P.J., Bhattacharjee, M.K., Kollia, E., DeSalle, R., Fine, D.H. & Figurski, D.H. (2000). Nonspecific adherence by *Actinobacillus actinomycetemcomitans* requires genes widespread in bacteria and archaea. *J. Bacteriol.* 182, 6169-6176.

Kachlany, S.C., Planet, P.J., DeSalle, R., Fine, D.H., & Figurski, D.H. (2001a). Genes for tight adherence of *Actinobacillus actinomycetemcomitans*: from plaque to plague to pond scum. *Trends Microbiol.* 9, 429-437.

Kachlany, S.C., Planet, P.J., DeSalle, R., Fine, D.H., Figurski, D.H. & Kaplan, J.B. (2001b). *flp-1*, first representative of a new pilin gene subfamily, is required for nonspecific adherence of *Actinobacillus actinomycetemcomitans*, *Mol. Microbiol.* 40, 542-554.

Kachlany, S.C. (2010). *Aggregatibacter actinomycetemcomitans* leukotoxin: from threat to therapy, *J. Dent. Res.* 89, 561-570.

Kaplan, J.B., Meyenhofer, M.F. & Fine, D.H. (2003). Biofilm growth and detachment of *Actinobacillus actinomycetemcomitans*, *J. Bacteriol.* 185, 1399-1404.

Kelk, P., Johansson, A., Claesson, R., Hänström, L. & Kalfas, S. (2003). Caspase 1 involvement in human monocyte lysis induced by *Actinobacillus actinomycetemcomitans* leukotoxin, *Infect. Immun.* 71, 4448-4455.

Kelk, P., Abd, H., Claesson, R., Sandström, G., Sjöstedt, A. & Johansson, A. (2011). Cellular and molecular response of human macrophages exposed to *Aggregatibacter actinomycetemcomitans* leukotoxin. *Cell Death Dis.* 2, e126.

Kovach, M.E., Phillips, R.W., Elzer, P.H., Roop II, R.M. & Peterson, K.M. (1994). pBBR1MCS: a broad-host-range cloning vector, *Biotechniques* 16, 800-802.

Kram, K.E., Hovel-Miner, G.A., Tomich, M. & Figurski, D.H. (2008). Transcriptional regulation of the *tad* locus in *Aggregatibacter actinomycetemcomitans*: a termination cascade, *J. Bacteriol.* 190, 3859-3868.

Komatsuzawa, H., Asakawa, R., Kawai, T., Ochiai, K., Fujiwara, T., Taubman, M.A., Ohara, M., Kurihara, H. & Sugai, M. (2002). Identification of six major outer membrane proteins from *Actinobacillus actinomycetemcomitans*, *Gene* 288, 195–201.

Labandeira-Rey, M., Brautigam, C.A. & Hansen, E.J. (2010). Characterization of the CpxRA regulon in *Haemophilus ducreyi*, *Infect. Immun.* 78, 4779-4791.

Lally, E.T., Hill, R.B., Kieba, I.R. & Korostoff, J. (1999). The interaction between RTX toxins and target cells, *Trends Microbiol.* 7, 356-361.

Löe, H. & Brown, L.J. (1991). Early onset periodontitis in the United States of America, *J. Periodontol.* 62, 608–616.

Lutkenhaus, J. & Sundaramoorthy, M. (2003). MinD and role of the deviant Walker A motif, dimerization and membrane binding in oscillation, *Mol. Microbiol.* 48, 295–303.

Marston, A.L. & Errington, J. (1999). Dynamic movement of the ParA-like Soj protein of *B. subtilis* and its dual role in nucleoid organization and developmental regulation, *Mol. Cell* 4, 673–682.

Mintz, K.P. (2004). Identification of an extracellular matrix protein adhesin, EmaA, which mediates the adhesion of *Actinobacillus actinomycetemcomitans* to collagen. *Microbiology* 150, 2677-2688.

O'Toole, G. A. & Kolter. R. (1998). Initiation of biofilm formation in *Pseudomonas fluorescens* WCS365 proceeds via multiple, convergent signalling pathways: a genetic analysis, *Mol. Microbiol.* 28, 449-461.

Pabo, C.O., Sauer, R.T., Sturtevant, J.M. & Ptashne, M. (1979). The lambda repressor contains two domains, *Proc. Natl. Acad. Sci. USA* 76, 1608-1612.

Paturel, L., Casalta, J.P., Habib, G., Nezri, M. & Raoult, D. (2004). *Actinobacillus actinomycetemcomitans* endocarditis, *Clin. Microbiol. Infect.* 10, 98–118.

Perez, B.A., Planet, P.J., Kachlany, S.C., Tomich, M., Fine, D.H. & Figurski, D.H. (2006). Genetic analysis of the requirement for *flp-2*, *tadV*, and *rcpB* in *Actinobacillus actinomycetemcomitans* biofilm formation. *J. Bacteriol.* 188, 6361-6375.

Perez-Cheeks, B.A., Planet, P.J., Sarkar, I.N., Clock, S.A., Xu, Q. & Figurski, D.H. (2012). The product of *tadZ*, a new member of the *parA/minD* superfamily, localizes to a pole in *Aggregatibacter actinomycetemcomitans*, *Mol. Microbiol.* 83, 694-711.

Pickett, C.L. & Whitehouse, C.A. (1999). The cytolethal distending toxin family, *Trends Microbiol.* 7, 292-297.

Planet, P.J., Kachlany, S.C., Fine, D.H., DeSalle, R., and Figurski, D.H. (2003) The Widespread Colonization Island (WCI) of *Actinobacillus actinomycetemcomitans*, *Nat. Genet.* 34, 193-198.

Quisel, J.D., Lin, D.C. & Grossman, A.D. (1999). Control of development by altered localization of a transcription factor in *B. subtilis*, *Mol. Cell* 4, 665–672.

Rahamat-Langendoen J.C., van Vonderen, M.G.A., Engström, L.J., Manson, W.L., van Winkelhoff, A.J. & Mooi-Kokenberg, E.A.N.M. (2011). Brain abscess associated with *Aggregatibacter actinomycetemcomitans*: case report and review of literature, *J. Clin. Periodontol.* 38, 702-706.

Ramirez-Arcos, S., Szeto, J., Dillon, J-A.R. & Margolin, W. (2002). Conservation of dynamic localization among MinD and MinE orthologues: oscillation of *Neisseria gonorrhoeae* proteins in *Escherichia coli*, *Mol. Microbiol.* 46, 493–504.

Ramsey, M.M. & Whiteley, M. (2009). Polymicrobial interactions stimulate resistance to host innate immunity through metabolite perception, *Proc. Natl. Acad. Sci. USA* 106, 1578-1583.

Raskin, D.M., & de Boer, P.A. (1999). Rapid pole-to-pole oscillation of a protein required for directing division to the middle of *Escherichia coli*, *Proc. Natl. Acad. Sci. USA* 96, 4971–4976.

Rose, J.E., Meyer, D.H. & Fives-Taylor, P.M. (2003). Aae, an autotransporter involved in adhesion of *Actinobacillus actinomycetemcomitans* to epithelial cells. *Infect. Immun.* 71, 2384-2393.

Rylev, M. &, Kilian, M. (2008). Prevalence and distribution of principal periodontal pathogens worldwide, *J. Clin. Periodontol.* 35, 346-361.

Schreiner, H.C., Sinatra, K., Kaplan, J.B., Furgang, D., Kachlany, S.C., Planet, P.J., Perez, B.A., Figurski, D.H. & Fine, D.H. (2003). Tight adherence genes of *Actinobacillus actinomycetemcomitans* are required for virulence in a rat model, *Proc. Natl. Acad. Sci. USA* 12, 7295-7300.

Slots, J. & Ting, M. (1999). *Actinobacillus actinomycetemcomitans* and *Porphyromonas gingivalis* in human periodontal disease: occurrence and treatment, *Periodontol. 2000* 20, 82-121.

Spinola, S. M., Fortney, K.R., Katz, B.P., Latimer, J.L., Mock, J.R., Vakevainen, M. & Hansen, E. J. (2003). *Haemophilus ducreyi* requires an intact *flp* gene cluster for virulence in humans, *Infect. Immun.* 71, 7178-7182.

Sternberg, N. & Hoess, R. (1983). The molecular genetics of bacteriophage P1, *Annu. Rev. Genet.* 17, 123-154.

Thomson, V.J., Bhattacharjee, M.K., Fine, D.H., Derbyshire, K.M. & Figurski, D.H. (1999). Direct selection of IS*903* transposon insertions by use of a broad host range vector: Isolation of catalase-deficient mutants of *Actinobacillus actinomycetemcomitans*, *J. Bacteriol.* 181, 7298-7307.

Tomich, M., Fine, D.H. & Figurski, D.H. (2006). The TadV protein of *Actinobacillus actinomycetemcomitans* is a novel aspartic acid prepilin peptidase required for maturation of the Flp1 pilin and TadE and TadF pseudopilins, *J. Bacteriol.* 188, 6899-6914.

Tomich, M., Planet, P.J. & Figurski, D.H. (2007). The *tad* locus: Postcards from the Widespread Colonization Island, *Nat. Rev. Microbiol.* 5, 363-375.

Tsai, C.C., McArthur, W.P., Baehni, P.C., Hammond, B.F. & Taichman, N.S. (1979). Extraction and partial characterization of a leukotoxin from a plaque-derived Gram-negative microorganism, *Infect. Immun.* 25, 427-439.

Van Winkelhoff, A. J. & Slots, J. (1999). *Actinobacillus actinomycetemcomitans* and *Porphyromonas gingivalis* in nonoral infections, *Periodontol. 2000* 20, 122-135.

Xu, Q., Christen, B., Chiu, H., Jaroszewski, L., Klock, H.E., Knuth, M.W., Miller, M.D., Elsliger, M., Deacon, A.M., Godzik, A., Lesley, S.A., Figurski, D.H., Shapiro, L. & Wilson, I.A. (2012). Structure of pilus assembly protein TadZ from *Eubacterium rectale*: implications for polar localization, *Mol. Microbiol.* 83, 712-727.

Yue, G., Kaplan, J.B., Furgang, D., Mansfield, K.G. & Fine, D.H. (2007). A second *Aggregatibacter actinomycetemcomitans* autotransporter adhesin exhibits specificity for buccal epithelial cells in humans and Old World primates, *Infect. Immun.* 75, 4440-4448.

Zambon, J.J. (1985). *Actinobacillus actinomycetemcomitans* in human periodontal disease, *J. Clin. Periodont.* 12, 1-20.

3

Site-Directed Mutagenesis and Yeast Reverse 2-Hybrid-Guided Selections to Investigate the Mechanism of Replication Termination

Deepak Bastia, S. Zzaman and Bidyut K. Mohanty
Department of Biochemistry and Molecular Biology,
Medical University of SC, Charleston, SC
USA

1. Introduction

DNA replication in prokaryotes, in budding yeast and in mammalian DNA viruses initiates from fixed origins (*ori*) and the replication forks are extended in either a bidirectional mode or in some cases unidirectionally (Cvetic and Walter, 2005; Sernova and Gelfand, 2008; Wang and Sugden, 2005; Weinreich et al., 2004). In higher eukaryotes there are preferred sequences located in AT-rich islands that serve as origins (Bell and Dutta, 2002). In many prokaryotes, the two replication forks initiated at *ori* on a circular chromosome meet each other at specific sequences called replication termini or *Ter* (Bastia and Mohanty, 1996; Kaplan and Bastia, 2009). The Ter sites bind to sequence-specific DNA binding proteins called replication terminator proteins that allow forks approaching from one direction to be impeded at the terminus, whereas forks coming from the opposite direction pass through the site unimpeded (Bastia and Mohanty, 1996, 2006; Kaplan and Bastia, 2009). Therefore, the mode of fork arrest is polar. The polarity of fork arrest in *Escherichia coli* and *Bacillus subtilis* is caused by the complexes of the terminator proteins called Tus and RTP (Replication Terminator Protein), respectively, with the cognate *Ter* sites to arrest the replicative helicase (such as DnaB in case of *E. coli*) in a polar mode (Kaul et al., 1994; Khatri et al., 1989; Lee et al., 1989; Sahoo et al., 1995). What is the mechanism of polar fork arrest and what might be the physiological functions of *Ter* sites? Using *E. coli* as the main example, with the aid of the techniques of site-directed mutagenesis, yeast reverse 2-hybrid based selection of random mutations (described below), and biochemical characterizations of the mutant forms of the Tus protein, many aspects of the mechanism of replication fork arrest at Tus-*Ter* complexes have been determined. This and a brief description of the current state of the knowledge of replication termination in eukaryotes have also been reviewed below.

Replication termini of *E. coli* **and the plasmid R6K:** Sequence-specific replication termini were first discovered in the drug resistance plasmid R6K (Crosa et al., 1976; Kolter and Helinski, 1978) and in its host *E. coli* (Kuempel et al., 1977). The terminus region of R6K was identified and sequenced (Bastia et al., 1981) and subsequently shown to consist of a pair of *Ter* sites with opposite polarity (Hidaka et al., 1988). An *in vitro* replication system was developed in which host cell extracts initiated replication of a plasmid DNA template and the moving forks were arrested at the *Ter* sites (Germino and Bastia, 1981). It was also suggested that a terminator protein that might cause fork arrest was likely to be host-encoded. Subsequently, the open reading frame (ORF) encoding the terminator protein was cloned and sequenced and the gene was named TUS (Terminus Utilizing Substance) (Hill et al., 1989). Tus protein was purified from cell extract of *E. coli* and shown to bind to the plasmid *Ter* sequences (Sista et al., 1991; Sista et al., 1989). The *TerC* region of *E. coli* was found to contain several *Ter* sites in two sets of 5 sites each with one cluster having the opposite polarity of fork arrest in comparison with that of the second set (Hill, 1992; Pelletier et al., 1988). Together, these sequences formed a replication trap (Fig.1A). For example, if the clockwise moving fork got arrested at *TerC*, it waited there for the counterclockwise fork to meet it at the site of arrest. The *Ter* consensus sequence is shown in Fig.1B. Site-directed mutagenesis showed the bases that are critical for Tus binding (Duggan et al., 1995; Sista et al., 1991). The complete process of initiation, elongation and termination has been carried out *in vitro* with 22 purified proteins that were necessary and sufficient for fork initiation, propagation and termination (Abhyankar et al., 2003).

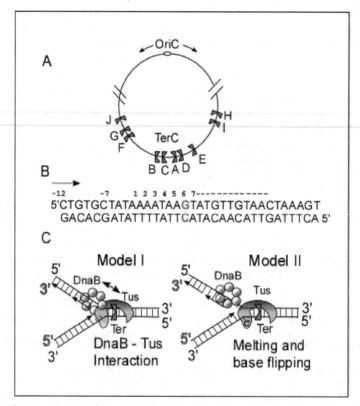

Fig. 1. Replication termini of *E. coli*. A, The bacterial replicon showing the origin and the *TerC* region at its antipode. The flat surfaces of the *Ter* sites indicate the permissive face and

the notched one the nonpermissive face; B, consensus Ter sequence showing the blocking
end at the left (arrow) and the nonblocking end at the right; the red C on the bottom strand
was reported to flip out upon Tus binding; C, two models of polar fork arrest. Model 1
postulates that both Tus binding to *Ter* and interaction or contact between the
nonpermissive face of the Tus-*Ter* complex with DnaB helicase causes polar arrest; model 2
suggests that it is strictly the Tus-*Ter* interaction and the partial melting of the DNA
catalyzed by DnaB and the flipping of C6 that causes strong affinity of Tus for *Ter*. The
helicase approaching the permissive face fails to induce high-affinity binding of Tus to *Ter*.

Using an *in vitro* helicase assay catalyzed by purified DnaB and Tus proteins, it was shown
that Tus binding to *Ter* acts as a polar contra- or anti-helicase and arrests helicase catalyzed
DNA unwinding in one orientation of the Tus-*Ter* complex while allowing the helicase to
pass through mostly unimpeded in the opposite orientation (Khatri et al., 1989; Lee et al.,
1989). It was also shown that the RTP of *B. subtilis* arrested *E. coli* DnaB helicase at the
cognate *Ter* sites of the Gram-positive bacterium *in vitro* was able to arrest DnaB of *E. coli* in
a polar mode. However, it did not arrest rolling circle replication of a plasmid (Kaul et al.,
1994). It is of some interest that not all helicases were arrested at Tus-*Ter* complexes because
helicases such as Rep and UvrD were not arrested by either orientations of Tus-*Ter* (Sahoo et
al., 1995). The Tus-*Ter* complex of *E. coli* could arrest forks with a very low efficiency *in vivo*
in the *B. subtils* host, as contrasted with their ability to arrest forks more efficiently in the
natural host. In addition to DnaB, RNA polymerase of bacteriophage T7 and *E. coli* were also
arrested in a polar mode, by the Tus-*Ter* complex (Mohanty et al., 1996, 1998). This latter
observation had raised the possibility that the Tus-*Ter* complex might just be a steric barrier
to unwinding because enzymes apparently as diverse as DnaB helicase and RNA
polymerases were arrested by the same complex. This mechanistic issue has been discussed
in more detail later.

Crystal structures of Terminator proteins: The first crystal structure of a terminator
apoprotein, namely that of RTP, showed that the protein was a symmetrical winged helix
dimer (Fig.2B) (Bussiere et al., 1995). The *Ter* sites of *B. subtilis* contain overlapping core and
auxiliary sequences with each site binding an RTP dimer (Hastings et al., 2005; Smith and
Wake, 1992; Wilce et al., 2001). How can a symmetrical protein arrest forks with polarity?
This question was subsequently answered when the crystal structure of two dimeric RTPs
bound to a complete bipartite *Ter* site was solved (Wilce et al., 2001). It was shown that the
structure of the protein-DNA complex is different at the core complex as contrasted with
that of the adjacent auxiliary complex. The crystal structure of Tus bound to *Ter* DNA
showed a bi-lobed protein with a positively charged cleft formed by several beta strands
that contacted the major groove of the DNA and distorted the latter from the canonical
structure (Fig.2A) (Kamada et al., 1996). The transverse view of Tus bound to a space-filling
model of DNA shows that the face that arrests replication forks and DnaB has a loop called
the L1 loop. The L1 loop appears to play a critical role in fork arrest.

Tus-DnaB interaction: We performed yeast 2-hybrid analysis (described below), confirmed
by *in vitro* affinity binding to immobilized Tus, to show that DnaB interacted with Tus
(Mulugu et al., 2001). The principles of forward 2-hybrid (Fields and Song, 1989) and reverse
2-hybrid analysis (Mulugu et al., 2001; Sharma et al., 2001) are shown in Fig.3. The open
reading frame (ORF) of a protein X is cloned in the correct reading frame to the
transcriptional activation domain of Gal4 of yeast (pGAD424-X). A suspected interacting

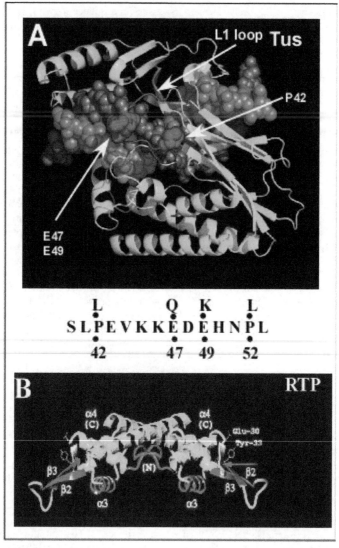

Fig. 2. Crystal structure of Tus-*Ter* complex of *E. coli* and RTP apoprotein of *B. subtilis*. A, crystal structure of Tus-*Ter* complex showing the blocking face with the L1 loop shown in red. Three residues, namely P42, E47 and E49, when mutated (see lower sequence) show impaired helicase arrest. P42L shows slightly reduced DNA binding; E47Q shows stronger DNA binding; and E49K shows no reduction in *Ter* binding but significant reduction in fork and helicase arrest. B, crystal structure of the RTP dimer apoprotein. The Tyr-33 arrow depicts a residue needed for the interaction of Tus with DnaB, as shown by a bifunctional labeled crosslinker that upon cleavage at an S-S bond transfers the label from RTP to DnaB.

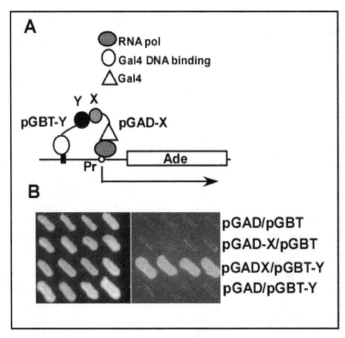

Fig. 3. Schematic representation of forward and reverse 2-hybrid selection. A, The plasmids
pGBT-Y and pGAD-X interact through interacting proteins X and Y and turn on the *Ade*
reporter gene leading to growth on adenine (ade) dropout minimal medium. Either X or Y is
mutagenized by low-fidelity PCR and introduced by transformation in the presence of the
other plasmid into the indicator yeast strain. B, X-Y interaction leads to growth on ade-
minus plates, and mutants that fail to interact show lack of growth on the selective plates.
Trivial mutations, *i.e.*, those containing deletions, nonsense mutations, or frame-shifts are
eliminated by Western blotting of cell extracts expressing the presumed X or Y mutant form.
Candidates are further characterized by functional and biochemical analyses.

protein Y is similarly fused in-frame to the ORF of the DNA binding domain of Gal4. The
yeast strain contains a transcriptional reporter (*Ade*) that is placed next to a promoter and
the binding site for the Gal4 DNA binding site. Neither pGAD424-X nor pGBT9-Y can
activate the transcription of the reporter gene. However, when both plasmids, each
containing a different marker (*e.g.*, *Leu* and *Trp*), are transformed into the reporter yeast
strain, X-Y interaction activates the reporter gene. Both plasmids are shuttle vectors that
contain an *ori* active in *E. coli* and also an *ori* (*ars*) of yeast. The transcription and translation
of the adenine (*Ade*) reporter causes the yeast cells to grow in an adenine dropout minimal
medium plate. The reverse 2-hybrid procedure was used to select for missense mutations
that break X-Y interaction as follows. Low fidelity PCR amplification of X (or Y) introduces
random mutations into the ORF. Then, for example, the mutagenized ORF of X in the
pGAD424 vector is used to transform the *Ade* reporter yeast strain containing a resident
pGBT9-Y plasmid. Colonies that have mutations that break X-Y interaction are initially
selected as clones growing on *Leu⁻Trp⁻* medium but failing to grow on *Leu⁻Trp⁻Ade⁻* dropout

plates. The mutations are expected to be a mixture of unwanted ones (*e.g.* missense, nonsense, frame-shifts) and useful ones (missense). The potential mutant clones are grown, cell-free lysates made and subjected to Western blots after polyacrylamide gel electrophoresis and developed with primary antibody raised against X followed by secondary reporter antibody. All clones that fail to produce the protein of the expected length are discarded, and those producing full length X-GAD are saved for further analysis.

Usually, the mutants are confirmed by co-immunoprecipitation of cell lysates with the anti-Y antibody (Ab) retained on agarose beads, stripping of the wild type (WT) X (or mutant X that should be in the wash), separation by gel electrophoresis and visualization with anti-Y Ab. Naturally, the authentic non-interaction mutant forms of X should no longer bind to Y or bind poorly. These "pull down" assays are used to confirm the reverse 2-hybrid results. If the interaction of X and Y is necessary for a biological function (*e.g.*, fork arrest at Tus-*Ter* complex), the X mutants that do not interact with protein Y are then tested by 2-dimensional agarose gel electrophoresis (Brewer and Fangman, 1987, 1988; Mohanty et al., 2006; Mohanty and Bastia, 2004) to determine whether they show the expected biochemical property (in this case, failure to arrest replication forks) (Mulugu et al., 2001). The reverse 2-hybrid approach is a powerful method that can yield mutants that specifically disrupt protein-protein interaction between a pair of known interacting proteins. This procedure can be followed up by isolation of additional mutations isolated by site-directed mutagenesis of residues close to the protein domain (as determined by X-ray crystallography) that contained the mutations recovered from the reverse 2-hybrid approach. A specific example is given below. By mutagenizing Tus by PCR, we were able to collect a pool of random mutants. We performed reverse 2-hybrid analysis of the mutant pool and recovered the mutation P42L (proline at position 42 to leucine) that fails to interact with DnaB. However, a P42L mutation also affected Tus-*Ter* binding to some extent. We mutagenized other residues by site-directed mutagenesis to isolate E47Q (glutamic acid at position 47 to glutamine) and E49K (glutamic acid at position 49 to lysine) (Fig. 2 and 3). Both of the latter mutants were defective in interaction with DnaB and in fork arrest *in vitro*. Whereas the E49K mutant form bound to *Ter* with the same affinity as WT Tus, E47Q had a higher DNA-binding affinity but was defective in fork arrest *in vivo* (Mulugu et al., 2001).

The yeast forward and reverse 2-hybrid analyses followed by biochemical analysis of Tus, showed that it contacted DnaB probably at the L1 loop because the only mutations that impaired helicase arrest and fork arrest without abolishing or significantly reducing Tus-*Ter* interaction were found only at the L1 loop. Another line of evidence for specific replisome-*Ter* interaction is inferred from the observation that that Tus-*Ter* complex works with very low efficiency when placed in *B. subtilis* cells as contrasted with their fork arrest efficiency in *E. coli in vivo* (Andersen et al., 2000).

If there is protein-protein interaction between Tus and DnaB and if this is necessary for fork arrest, how does Tus also promote polar arrest of RNA polymerase, an enzyme apparently different in structure from DnaB? One possible explanation is that Tus might make an equivalent contact with RNA polymerase to inhibit its progression, or else a different mechanism could be operating here. It should, however, be clearly stated that this line of reasoning does not necessarily disprove the first explanation. Based on the data discussed above, we have suggested a model of fork arrest that involves not only stable Tus-*Ter* interaction, but also protein-protein contacts between the DnaB helicase and the L1 loop of Tus (Fig.1C and Fig.2).

Base flipping and DNA melting: An alternative explanation of polar arrest is suggested in model 2 (Fig.1C). X-ray crystallography of Tus bound to linear DNA had shown all Watson-Crick base pairing (Kamada et al., 1996). However, it was reported that a forked DNA that had single stranded regions when co-crystallized with Tus showed a flipped base (C6 in Fig 1C, model 2). It was suggested that both DNA melting and base flipping and the capture of the flipped base by Tus greatly enhanced Tus binding for *Ter* when the helicase approached the blocking end of the Tus-*Ter* complex. The enzyme, when approaching the complex from the non-blocking end, displaced Tus from *Ter*. This interpretation was based on binding studies of Tus to *Ter* on partially single-stranded DNA having a flipped C (Mulcair et al., 2006). Unfortunately, these binding studies were performed between 150 mM-250 mM KCl at which DNA replication and DnaB activity *in vitro* is inhibited by >90% . Curiously, when binding was performed closer to a physiological salt concentration that is permissive of DNA replication, this high binding affinity was greatly reduced to that of the interaction between linear double stranded Ter DNA and Tus (Kaplan and Bastia, 2009). It was therefore necessary to carefully test model 2 to determine its authenticity.

An Independent test of the melting-flipping model shows that it is unnecessary for polar fork arrest: We wished to rigorously test model 2, which postulated that DNA melting and base flipping together could explain polar fork arrest under a physiological salt concentration that permitted DNA replication to occur (Bastia et al., 2008). We reasoned that the model could be tested if one could temporally and spatially separate DNA unwinding by DnaB helicase from its ATP-dependent locomotion on DNA (double- or single-stranded). It is known that when encountering a linear DNA with a 5' tail and 3' blunt end, DnaB enters DNA with both strands passing through the central channel of DnaB (Kaplan, 2000). The translocation of DnaB on double-stranded DNA (dsDNA) requires ATP hydrolysis. We constructed the DNA substrate shown in Fig. 4. The DnaB helicase enters the substrate from the left by riding the 5'-single-stranded tail, slides over dsDNA containing a *Ter* site present in both orientations and upon reaching the forked structure with a 3' overhang, DnaB unwinds this labeled strand (shown in blue). In the blocking orientation of Tus-*Ter* complex, the DnaB helicase slides on the dsDNA until it reached the *Ter* site, at which it is arrested, as shown by its failure to melt off the labeled 3' tail shown in blue. In the reverse orientation of Tus-*Ter*, the DnaB sliding should displace Tus from *Ter* and continue sliding until it reached the 3' overhang fork-like structure. At this point it should melt the labeled oligonucleotide, causing its release that can be resolved in a polyacrylamide gel at neutral pH and quantified (Fig.4). Our experiments showed that DnaB sliding, that involved no melting of DNA, not even a transient one, was arrested in a polar mode at a Tus-*Ter* complex. We proceeded to confirm the results further by introducing a pair of site-directed A-T inter-strand cross-links at two residues preceding C6. This covalent interstrand linkage prevented any chance of even transient DNA melting catalyzed by DnaB preceding the C6 residue. We confirmed that in such a substrate, DnaB sliding was arrested in a polar mode by the Tus-*Ter* complex only when present in the blocking orientation. These experiments led us to conclude that under physiological conditions a melting-flipping mechanism is not necessary (and probably does not occur) to cause polar fork arrest (Bastia et al., 2008).

Resolution of daughter DNA molecules at Ter sites: Following fork arrest at *Ter* sites, the daughter DNA molecules are resolved by a special type II topoisomerase, namely Topo IV (Espeli et al., 2003). It has been reported that this topoisomerase is stimulated by the actin-like MreB protein that acts near the resolution site *dif* that resolves dimers generated by recombination (Madabhushi and Marians, 2009).

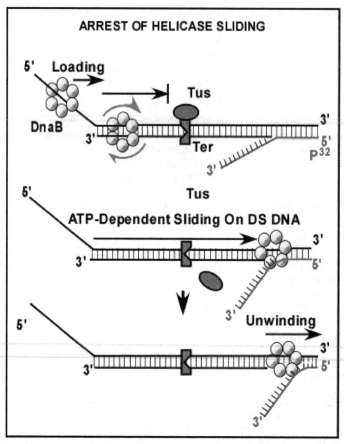

Fig. 4. A substrate designed to separate temporally and spatially DnaB translocation from DNA unwinding. A 5′ tailed DNA with otherwise a blunt end on the complementary strand enters the substrate and then slides over the dsDNA until it meets the fork like structure (in blue) and unwinds the labeled strand. If a Tus-*Ter* complex is present in a blocking orientation, the sliding DnaB is arrested, thereby preventing the unwinding of the blue strand; a *Ter* site in the permissive orientation when bound to Tus displaces Tus and slides down the substrate and unwinds the blue strand. The results showed that DnaB sliding, without any DNA melting was arrested in a polar mode by the Tus-*Ter* complex, thereby showing that DNA unwinding (and presumably base flipping) is not necessary for polar helicase/ fork arrest.

Replication termini in eukaryotes: Many, perhaps all, eukaryotes have sequence-specific replication termini located in their ribosomal DNA (rDNA) array. For example, *Saccharomyces cerevisiae* contains a pair of Ter sites in one of the nontranscribed spacers of each rDNA unit between the sequences encoding the 35S RNA and the 5S RNA (Brewer and Fangman, 1988; Brewer et al., 1992; Ward et al., 2000). The second spacer contains a replication *ori* (*ars*; see Fig.5). The Ter sites bind to the replication terminator protein called Fob1 (fork blockage) (Kobayashi, 2003; Kobayashi and Horiuchi, 1996; Mohanty and Bastia, 2004). The Fob1 protein bound to Ter sites prevents replication forks moving from right to left from colliding with the strong transcription of 35S RNA. It has been shown that transcription-replication collision causes not only fork stalling but also stalled RNA polymerase and an incomplete RNA transcript that can hybridize with DNA to form an R loop. R loops, especially the single stranded DNA therein, is susceptible to physical and enzymatic damage *in vivo* which causes genome instability (Helmrich et al., 2011).

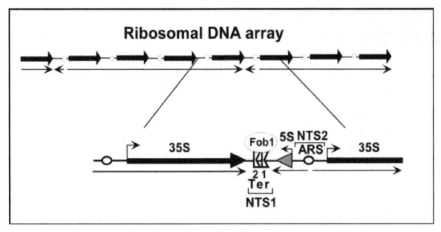

Fig. 5. rDNA repeat region in chromosome XII of *S. cerevisiae* showing the location of the two Ter sites in the nontranscribed spacer 1 (NTS1). The replication is initiated bidirectionally from the *ars* present in nontranscribed spacer 2 (NTS2). The Ter sites prevent replication forks moving to the left from the *ars* from running into RNA polymerase transcribing the 35S rRNA precursor.

The Fob1 protein is multifunctional and loads histone deacetylase to silence intra-chromatid recombination in the tandem array of ~200 rDNA repeats that might otherwise lead to unscheduled loss or gain of rDNA repeats (Bairwa et al., 2010; Huang et al., 2006; Huang and Moazed, 2003). Fob1 protein is also a transcriptional activator and controls exit from mitosis (Bastia and Mohanty, 2006; Stegmeier et al., 2004).

One of the facile techniques to study Fob1 function is to perform segment-directed mutagenesis, which is shown schematically (Fig.6). A segment of an ORF flanked by regions of homology (also from the ORF) is amplified by PCR under conditions of low fidelity synthesis in which one of the dNTPs is present at a suboptimal concentration. This leads to misincorporation of the base into DNA causing random mutations. A plasmid containing a gap corresponding to the segment being mutagenized and the PCR products are used to transform yeast. The mutagenized DNA segment gets incorporated into the plasmid by gap repair caused by the homologous recombination machinery of yeast with high efficiency, thus generating a pool of potential mutants contained in the plasmid. The plasmid contains a marker expressed in yeast (*e.g., Leu*) and an *ars*. Using this protocol, we extensively mutagenized Fob1 and were able to identify many of its functional domains, such as its

DNA binding domain and a domain for its interaction with the silencing linker protein called Net1. Net1 recruits the histone deacetylase Sir2 onto Fob1 by direct protein-protein interaction between Net1 and Sir2 on one hand and between Net1 and Fob1 on the other, and loads Sir2 near the Ter sites. This process, as noted above, causes silencing of rDNA and prevents unwanted recombination (Bairwa et al., 2010; Mohanty and Bastia, 2004). At this time, the detailed mechanism of replication termination in eukaryotes has not been elucidated. However, it is known that two intra-S checkpoint proteins called Tof1 and its interacting partner called Csm3 are necessary for stable fork arrest at Ter because the Tof1-Csm3 complex protects the Fob1 protein from getting displaced from the Ter site by the action of the helicase Rrm3 (Mohanty et al., 2006, 2009). The catenated daughter molecules at Ter sites in *S. cerevisiae* are separated from each other by Topo II (Baxter and Diffley, 2008; Fachinetti et al., 2010).

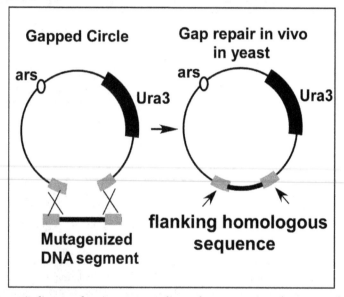

Fig. 6. Schematic diagram showing segment-directed mutagenesis and recovery of mutants by gap repair. The gapped plasmid is prepared by restriction site cutting inside the ORF. The DNA segment is mutagenized by low-fidelity PCR that includes primers with homologous flanking sequence. Transformation of a mixture of mutagenized DNA mixed with the gapped plasmid results in a pool of plasmids, some of which should have random base changes within the mutagenized DNA segment

We have recently reported that the Reb1 terminator protein binding to 2 Ter sites of fission yeast act in a cooperative fashion. The dimeric Reb1 protein, for example, brings into contact a Ter site located on chromosome 2 with two Ter sites located on chromosome 1. Interestingly there was no interaction observed between sites on chromosome 1 and 2 with the Ter sites located in the two rDNA clusters present on chromosome 3. It seems that the Ter-Ter interactions are not random. We further reported that the interactions called "chromosome kissing' modulated the activities of the Ter sites (Singh et al., 2010).

Physiological function of the replication termini: In prokaryotes, the replication termini perform at least 2 functions: (i) these serve as a replication trap and confine the meeting of the two approaching forks to the *TerC* region (Fig.1) where the dimer resolution (*dif*) sites are located. This activity probably facilitates chromosome segregation (Wake, 1997); and (ii) the terminus, in plasmid chromosomes prevents accidental switch to a rolling circle mode of replication that would generate unwanted linearly catenated chromosome (Dasgupta et al., 1991). In eukaryotes, the termini probably serve as barriers to transcription-replication collision that might generate destabilizing R loops. The termini are also known to be involved in cellular differentiation of fission yeast (Dalgaard and Klar, 2000, 2001). As noted above, Fob1 protein has diverse other functions (Bastia and Mohanty, 2006; Kaplan and Bastia, 2009).

In summary, replication termination at site-specific termini is an important part of DNA replication that invites further investigation, especially in eukaryotes, because of its role in various DNA transactions including maintenance of genome stability.

Acknowledgement: We thank Dr. G. Krings and other members of our group for their valuable contributions to the investigations of replication termination. Our work was supported by a grant from the NIGMS.

2. References

Abhyankar, M.M., Zzaman, S., and Bastia, D. (2003). Reconstitution of R6K DNA replication in vitro using 22 purified proteins. J Biol Chem *278*, 45476-45484.

Andersen, P.A., Griffiths, A.A., Duggin, I.G., and Wake, R.G. (2000). Functional specificity of the replication fork-arrest complexes of Bacillus subtilis and Escherichia coli: significant specificity for Tus-Ter functioning in E. coli. Mol Microbiol *36*, 1327-1335.

Bairwa, N.K., Zzaman, S., Mohanty, B.K., and Bastia, D. (2010). Replication fork arrest and rDNA silencing are two independent and separable functions of the replication terminator protein Fob1 of Saccharomyces cerevisiae. J Biol Chem *285*, 12612-12619.

Bastia, D., Germino, J., Crosa, J.H., and Ram, J. (1981). The nucleotide sequence surrounding the replication terminus of R6K. Proc Natl Acad Sci U S A *78*, 2095-2099.

Bastia, D., and Mohanty, B.K. (1996). Mechanisms for completing DNA replication. DNA Replication in Eukaryotic Cells (M DePamphilis, Ed) Cold Spring Harbor Laboratory Press, NY, 177-215.

Bastia, D., and Mohanty, B.K. (2006). Termination of DNA Replication. DNA replication and human disease (ed ML DePamphilis), Cold Spring Harbor Laboratory Press, Cold Spring Harbor, New York, 155-174.

Bastia, D., Zzaman, S., Krings, G., Saxena, M., Peng, X., and Greenberg, M.M. (2008). Replication termination mechanism as revealed by Tus-mediated polar arrest of a sliding helicase. Proc Natl Acad Sci U S A *105*, 12831-12836.

Baxter, J., and Diffley, J.F. (2008). Topoisomerase II inactivation prevents the completion of DNA replication in budding yeast. Mol Cell *30*, 790-802.

Bell, S.P., and Dutta, A. (2002). DNA replication in eukaryotic cells. Annu Rev Biochem *71*, 333-374.

Brewer, B.J., and Fangman, W.L. (1987). The localization of replication origins on ARS plasmids in S. cerevisiae. Cell *51*, 463-471.

Brewer, B.J., and Fangman, W.L. (1988). A replication fork barrier at the 3' end of yeast ribosomal RNA genes. Cell 55, 637-643.

Brewer, B.J., Lockshon, D., and Fangman, W.L. (1992). The arrest of replication forks in the rDNA of yeast occurs independently of transcription. Cell 71, 267-276.

Bussiere, D.E., Bastia, D., and White, S.W. (1995). Crystal structure of the replication terminator protein from B. subtilis at 2.6 A. Cell 80, 651-660.

Crosa, J.H., Luttropp, L.K., and Falkow, S. (1976). Mode of replication of the conjugative R-plasmid RSF1040 in Escherichia coli. J Bacteriol 126, 454-466.

Cvetic, C., and Walter, J.C. (2005). Eukaryotic origins of DNA replication: could you please be more specific? Semin Cell Dev Biol 16, 343-353.

Dalgaard, J.Z., and Klar, A.J. (2000). swi1 and swi3 perform imprinting, pausing, and termination of DNA replication in S. pombe. Cell 102, 745-751.

Dalgaard, J.Z., and Klar, A.J. (2001). A DNA replication-arrest site RTS1 regulates imprinting by determining the direction of replication at mat1 in S. pombe. Genes Dev 15, 2060-2068.

Dasgupta, S., Bernander, R., and Nordstrom, K. (1991). In vivo effect of the tus mutation on cell division in an Escherichia coli strain where chromosome replication is under the control of plasmid R1. Res Microbiol 142, 177-180.

Duggan, L.J., Hill, T.M., Wu, S., Garrison, K., Zhang, X., and Gottlieb, P.A. (1995). Using modified nucleotides to map the DNA determinants of the Tus-TerB complex, the protein-DNA interaction associated with termination of replication in Escherichia coli. J Biol Chem 270, 28049-28054.

Espeli, O., Levine, C., Hassing, H., and Marians, K.J. (2003). Temporal regulation of topoisomerase IV activity in E. coli. Mol Cell 11, 189-201.

Fachinetti, D., Bermejo, R., Cocito, A., Minardi, S., Katou, Y., Kanoh, Y., Shirahige, K., Azvolinsky, A., Zakian, V.A., and Foiani, M. (2010). Replication termination at eukaryotic chromosomes is mediated by Top2 and occurs at genomic loci containing pausing elements. Mol Cell 39, 595-605.

Fields, S., and Song, O. (1989). A novel genetic system to detect protein-protein interactions. Nature 340, 245-246.

Germino, J., and Bastia, D. (1981). Termination of DNA replication in vitro at a sequence-specific replication terminus. Cell 23, 681-687.

Hastings, A.F., Otting, G., Folmer, R.H., Duggin, I.G., Wake, R.G., Wilce, M.C., and Wilce, J.A. (2005). Interaction of the replication terminator protein of Bacillus subtilis with DNA probed by NMR spectroscopy. Biochem Biophys Res Commun 335, 361-366.

Helmrich, A., Ballarino, M., and Tora, L. (2011). Collisions between replication and transcription complexes cause common fragile site instability at the longest human genes. Mol Cell 44, 966-977.

Hidaka, M., Akiyama, M., and Horiuchi, T. (1988). A consensus sequence of three DNA replication terminus sites on the E. coli chromosome is highly homologous to the terR sites of the R6K plasmid. Cell 55, 467-475.

Hill, T.M. (1992). Arrest of bacterial DNA replication. Annu Rev Microbiol 46, 603-633.

Hill, T.M., Tecklenburg, M.L., Pelletier, A.J., and Kuempel, P.L. (1989). tus, the trans-acting gene required for termination of DNA replication in Escherichia coli, encodes a DNA-binding protein. Proc Natl Acad Sci U S A 86, 1593-1597.

Huang, J., Brito, I.L., Villen, J., Gygi, S.P., Amon, A., and Moazed, D. (2006). Inhibition of homologous recombination by a cohesin-associated clamp complex recruited to the rDNA recombination enhancer. Genes Dev 20, 2887-2901.

Huang, J., and Moazed, D. (2003). Association of the RENT complex with nontranscribed and coding regions of rDNA and a regional requirement for the replication fork block protein Fob1 in rDNA silencing. Genes Dev 17, 2162-2176.

Kamada, K., Horiuchi, T., Ohsumi, K., Shimamoto, N., and Morikawa, K. (1996). Structure of a replication-terminator protein complexed with DNA. Nature 383, 598-603.

Kaplan, D.L. (2000). The 3'-tail of a forked-duplex sterically determines whether one or two DNA strands pass through the central channel of a replication-fork helicase. J Mol Biol 301, 285-299.

Kaplan, D.L., and Bastia, D. (2009). Mechanisms of polar arrest of a replication fork. Mol Microbiol 72, 279-285.

Kaul, S., Mohanty, B.K., Sahoo, T., Patel, I., Khan, S.A., and Bastia, D. (1994). The replication terminator protein of the gram-positive bacterium Bacillus subtilis functions as a polar contrahelicase in gram-negative Escherichia coli. Proc Natl Acad Sci U S A 91, 11143-11147.

Khatri, G.S., MacAllister, T., Sista, P.R., and Bastia, D. (1989). The replication terminator protein of E. coli is a DNA sequence-specific contra-helicase. Cell 59, 667-674.

Kobayashi, T. (2003). The replication fork barrier site forms a unique structure with Fob1p and inhibits the replication fork. Mol Cell Biol 23, 9178-9188.

Kobayashi, T., and Horiuchi, T. (1996). A yeast gene product, Fob1 protein, required for both replication fork blocking and recombinational hotspot activities. Genes Cells 1, 465-474.

Kolter, R., and Helinski, D.R. (1978). Activity of the replication terminus of plasmid R6K in hybrid replicons in Escherichia coli. J Mol Biol 124, 425-441.

Kuempel, P.L., Duerr, S.A., and Seeley, N.R. (1977). Terminus region of the chromosome in Escherichia coli inhibits replication forks. Proc Natl Acad Sci U S A 74, 3927-3931.

Lee, E.H., Kornberg, A., Hidaka, M., Kobayashi, T., and Horiuchi, T. (1989). Escherichia coli replication termination protein impedes the action of helicases. Proc Natl Acad Sci U S A 86, 9104-9108.

Madabhushi, R., and Marians, K.J. (2009). Actin homolog MreB affects chromosome segregation by regulating topoisomerase IV in Escherichia coli. Mol Cell 33, 171-180.

Mohanty, B.K., Bairwa, N.K., and Bastia, D. (2006). The Tof1p-Csm3p protein complex counteracts the Rrm3p helicase to control replication termination of Saccharomyces cerevisiae. Proc Natl Acad Sci U S A 103, 897-902.

Mohanty, B.K., Bairwa, N.K., and Bastia, D. (2009). Contrasting Roles of Checkpoint Proteins as Recombination Modulators At Fob1-Ter Complexes With or Without Fork Arrest. Eukaryot Cell 8, 487-495.

Mohanty, B.K., and Bastia, D. (2004). Binding of the replication terminator protein Fob1p to the Ter sites of yeast causes polar fork arrest. J Biol Chem 279, 1932-1941.

Mohanty, B.K., Sahoo, T., and Bastia, D. (1996). The relationship between sequence-specific termination of DNA replication and transcription. EMBO J 15, 2530-2539.

Mohanty, B.K., Sahoo, T., and Bastia, D. (1998). Mechanistic studies on the impact of transcription on sequence-specific termination of DNA replication and vice versa. J Biol Chem 273, 3051-3059.

Mulcair, M.D., Schaffer, P. M., Oakly, A.J.Cross, H.F., Neylon, C., Hill, T.M. and Dixon, N. (2006) Cell, 125: 1309-1313.

Mulugu, S., Potnis, A., Shamsuzzaman, Taylor, J., Alexander, K., and Bastia, D. (2001). Mechanism of termination of DNA replication of Escherichia coli involves helicase-contrahelicase interaction. Proc Natl Acad Sci U S A 98, 9569-9574.

Pelletier, A.J., Hill, T.M., and Kuempel, P.L. (1988). Location of sites that inhibit progression of replication forks in the terminus region of Escherichia coli. J Bacteriol 170, 4293-4298.

Sahoo, T., Mohanty, B.K., Lobert, M., Manna, A.C., and Bastia, D. (1995). The contrahelicase activities of the replication terminator proteins of Escherichia coli and Bacillus subtilis are helicase-specific and impede both helicase translocation and authentic DNA unwinding. J Biol Chem 270, 29138-29144.

Sernova, N.V., and Gelfand, M.S. (2008). Identification of replication origins in prokaryotic genomes. Brief Bioinform 9, 376-391.

Sharma, R., Kachroo, A., and Bastia, D. (2001). Mechanistic aspects of DnaA-RepA interaction as revealed by yeast forward and reverse two-hybrid analysis. EMBO J 20, 4577-4587.

Singh, S.K., Sabatinos, S., Forsburg, S., and Bastia, D. (2010). Regulation of replication termination by Reb1 protein-mediated action at a distance. Cell 142, 868-878.

Sista, P.R., Hutchinson, C.A., 3rd, and Bastia, D. (1991). DNA-protein interaction at the replication termini of plasmid R6K. Genes Dev 5, 74-82.

Sista, P.R., Mukherjee, S., Patel, P., Khatri, G.S., and Bastia, D. (1989). A host-encoded DNA-binding protein promotes termination of plasmid replication at a sequence-specific replication terminus. Proc Natl Acad Sci U S A 86, 3026-3030.

Smith, M.T., and Wake, R.G. (1992). Definition and polarity of action of DNA replication terminators in Bacillus subtilis. J Mol Biol 227, 648-657.

Stegmeier, F., Huang, J., Rahal, R., Zmolik, J., Moazed, D., and Amon, A. (2004). The replication fork block protein Fob1 functions as a negative regulator of the FEAR network. Curr Biol 14, 467-480.

Wake, R.G. (1997). Replication fork arrest and termination of chromosome replication in Bacillus subtilis. FEMS Microbiol Lett 153, 247-254.

Wang, J., and Sugden, B. (2005). Origins of bidirectional replication of Epstein-Barr virus: models for understanding mammalian origins of DNA synthesis. J Cell Biochem 94, 247-256.

Ward, T.R., Hoang, M.L., Prusty, R., Lau, C.K., Keil, R.L., Fangman, W.L., and Brewer, B.J. (2000). Ribosomal DNA replication fork barrier and HOT1 recombination hot spot: shared sequences but independent activities. Mol Cell Biol 20, 4948-4957.

Weinreich, M., Palacios DeBeer, M.A., and Fox, C.A. (2004). The activities of eukaryotic replication origins in chromatin. Biochim Biophys Acta 1677, 142-157.

Wilce, J.A., Vivian, J.P., Hastings, A.F., Otting, G., Folmer, R.H., Duggin, I.G., Wake, R.G., and Wilce, M.C. (2001). Structure of the RTP-DNA complex and the mechanism of polar replication fork arrest. Nat Struct Biol 8, 206-210.

4

Site-Directed Mutagenesis as a Tool to Characterize Specificity in Thiol-Based Redox Interactions Between Proteins and Substrates

Luis Eduardo S. Netto[1] and Marcos Antonio Oliveira[2]
[1]*Instituto de Biociências – Universidade de Sao Paulo*
[2]*Universidade Estadual Paulista – Campus do Litoral Paulista*
Brazil

1. Introduction

Redox pathways are involved in several processes in biology, such as signal transduction, regulation of gene expression, oxidative stress and energy metabolism. Proteins are the central mediators of electron transfer processes. Many of these proteins rely on non-proteinaceous redox cofactors (such as NAD^+; FAD; heme; or Cu, Fe or other transition metals) for their redox activity. In contrast, other proteins use cysteine residues for this property (Netto et al., 2007). The amino acid cysteine has low reactivity for redox transitions (Winterbourn and Metodiewa, 1999; Wood et al., 2003; Marino and Gladishev, 2011). However, protein folding can generate environments in which cysteine residues are reactive. Examples are reduction (or isomerization or formation) of disulfide bonds, reduction of methionine thioesther-sulfoxide, degradation of peptide bonds, peroxide reduction, and others (Lindahl et al., 2011).

Glutathione (GSH) is by far the major non-proteinaceous thiol in cells that plays a central role in several redox processes, such as xenobiotic excretion and antioxidant defense. GSH is composed of three amino acids: glutamate, cysteine and glycine. GSH synthesis is performed in two steps, which occur mainly in liver cells. In the first step, the γ-glutamylcysteine synthetase enzyme catalyses the rate-limiting step, with the formation of an unusual peptide bond between the gamma-carboxyl group of the side chain of glutamate and the primary amino group of a cysteine in an ATP-dependent reaction. Then GSH synthetase catalyzes the formation of a peptide bond between the carboxyl group of cysteine (from the dipeptide γ-glutamylcysteine) with the amino group of a glycine. This tripeptide is considered to be the major redox buffer in mammalian cells, and it is a substrate of two relevant groups of enzymes: glutathione transferases and glutathione peroxidases. It is thought that most healthy cells have higher GSH/GSSG (glutathione disulfide) ratios than sick ones (Berndt et al., 2007; Jacob et al., 2003; Jones, 2006).

Besides glutathione, there is a high number of Cys-based (see Table 1 in the chapter by Figurski et al. for amino acid abbreviations) redox proteins. These Cys-based proteins are very versatile. The oxidation states of their sulfur atoms can vary from +6 to -2 (Jacob et al., 2003). One of the most widespread functions of Cys-based proteins is the catalysis of thiol-disulfide exchange reactions, by which these enzymes control the oxidation state (dithiol or disulfide) in their targets/substrates (Netto et al., 2007). Thioredoxins and glutaredoxins (also known as thioltransferases) are disulfide reductases, whereas protein disulfide isomerases are also involved in the oxidation of dithiols and/or the shuffling of disulfides. Furthermore, Cys-based proteins can also control the levels of other eletrophiles, such as peroxides (in the cases of peroxiredoxins and GSH peroxidases), xenobiotics (GSH transferases) and sulfoxides (methionine sulfoxide reductases). Therefore, this large repertoire of proteins, together with GSH, is part of a complex network that, in a dynamic fashion, controls intracellular redox balance.

The classical view is that the reactivity of a cysteine sulfhydryl group is related to its pK_a, since its deprotonated form (thiolate = RS⁻) is more nucleophilic and, therefore, reacts faster than the equivalent protonated form (R-SH). According to this view, the lower the pK_a of a thiol, the higher will be the availability of the more nucleophilic species, the thiolate. The sulfhydryl groups of most cysteines (either linked to a polypeptide backbone or the free amino acid) possess low reactivity, which has been related to the fact that their pK_a values are around 8.5 (Benesch and Benesch, 1955). In contrast, most redox proteins possess a reactive cysteine that is stabilized in the thiolate form by a basic residue - in most cases by a lysine, histidine or arginine residue (Copley et al., 2004).

However, a decrease in the pK_a value of several orders of magnitude would give rise to an increase in thiolate concentration, with a maximum increase of one order of magnitude (Ferrer-Sueta et al., 2011). However, as an example of Cys-based redox proteins, peroxiredoxins reacts one to ten million times faster with peroxides than the corresponding reaction with the free amino acid cysteine (Winterbourn and Hampton, 2008). Therefore, factors other than thiolate availability should be taken into account. Indeed the stabilization of the transition state by active-site residues was recently proposed to be the catalytic power of peroxiredoxins. Site-directed mutagenesis was employed to test these hypotheses (Hall et al., 2010; Nagy et al., 2011).

It is clear that Cys-based proteins present reactive Cys residues to specific reactions, most of them being the nucleophilic substitution (S_N2) type. Indeed peroxiredoxins are effective in reducing peroxides; but they are poor in reducing other eletrophiles, such as chroloamines (Peskin et al., 2007). In contrast, glutaredoxins are powerful GSH-dependent disulfide reductants. In spite of the fact that their reactive Cys residues have low pK_a values (<4.0), these oxido-reductases are unable to reduce O-O bonds (Discola et al., 2009).

In line with the observation that Cys-based circuits display high specificity, a new concept of oxidative stress was proposed by Jones (2006). Since several antioxidant interventions failed to have therapeutic effects, it was thought that oxidative stress leads to alterations of discrete pathways, rather than to an overall redox imbalance. Therefore, perhaps an antioxidant intervention would be more effective if it were directed to specific pathways, i.e., the oxidative stress would be better defined as a disruption of a specific pathway (Jones, 2006). For instance, some signal transduction pathways are activated by oxidized, but not by reduced, thioredoxin (Trx) (Berndt et al., 2007), e.g., only reduced Trx1 binds Ask-1, thereby inhibiting the kinase activity of Ask-1, whereas oxidation of Trx1 leads to dissociation of the complex and activation of Ask-1, which can trigger apoptosis (Saitoh et al., 1998). Another example is the activation of NF-κB. Binding of subunit p50 to its target sequence in DNA requires the reduction of a single cysteinyl residue in the nucleus by Trx1 (Matthews et al.,

1992; Hayashi et al., 1993). Also circadian cycles depend on Cys-based redox signaling (O'Neill et al., 2011). The specificity of these pathways involves protein–protein interactions. The identification of the amino acids involved is relevant for the comprehension of patho-physiological phenomena.

2. Approach

Our group has followed an approach for studying Cys-based redox systems that involves multiple methodologies. Initially we decided to study yeast thiol-based systems, such as the thioredoxin and glutaredoxin ones. Yeast is a convenient system because it is very amenable to genetic manipulation. We obtained from *EUROSCARF* (http://web.unifrankfurt.de/fb15/mikro/euroscarf/complete.html) a collection of about four thousand strains, each one with a single deletion of a specific gene. In the case of the cytosolic thioredoxin system from *Saccharomyces cerevisiae*, we have elucidated the three-dimensional structures of all proteins by NMR (*i.e.*, ScTrx1 and ScTrx2), in collaboration with the group of Dr. Almeida (Pinheiro et al., 2008; Amorim et al., 2007), as well as by crystallography, *i.e.*, ScTrxR1, also known as Trr1 (Oliveira et al., 2010). We have also elucidated structures of the yeast glutaredoxins (Discola et al., 2009). With this information, together with available public structural, biochemical, and enzymatic data, we were able to generate a hypothesis about the mechanistic aspects of these redox pathways that could be tested by site-directed mutagenesis.

Recently we have done this approach with the bacterial thiol-based systems, as a consequence of our participation in the genome sequencing project for the phytopathogen *Xylella fastidiosa* (Simpson et al., 2000). Again site-directed mutagenesis was employed to test a hypothesis generated by experimental work. The hypothesis proposed the involvement of amino acid residues in catalysis. The interpretation of the data was not always straightforward, probably because amino acids can interact with other residues in a protein to give unpredictable effects. In the following sections, we will describe what we have learned in different thiol-based systems, using site-directed mutagenesis to test hypotheses.

3. Characterization of Cys-based proteins

3. 1. Cys-based proteins from *Saccharomyces cerevisiae*

3.1.1 Molecular aspects of specific redox protein-protein interactions in the cytosolic thioredoxin system

Thioredoxin appears to be an ancient protein, since it is widespread among all living organisms. These small proteins (12–13 kDa) possess disulfide reductase activity endowed by two vicinal cysteines present in a CXXC residue motif - typically CGPC. The cysteines are used to reduce target proteins that are recognized by other domains of thioredoxin polypeptide. The reduction of target proteins results in a disulfide bridge between the two cysteines from the thioredoxin CXXC motif, which is then reduced by thioredoxin reductase using reducing equivalents from NADPH. Some of the target proteins of thioredoxin include ribonucleotide reductase (important for DNA synthesis), methionine sulfoxide reductase, peroxiredoxins and transcription factors such as p53 and NF-kB (reviewed by Powis and Montfort, 2001). Therefore, the thioredoxin system is composed of NADPH, thioredoxin reductase and thioredoxin.

Proteins endowed with thioredoxin reductase activity are also widespread and comprise enzymes with different redox centers. Thioredoxin reductase enzymes belong to the nucleotide pyridine disulfide oxidoreductase family, which includes glutathione reductase, alkyl hydroperoxide reductase F (AhpF), and lipoamide dehydrogenase (Williams et al., 2000). Constituents of this family are homodimeric flavoproteins that also contain one or two dithiol-disulfide motifs - CXXXC and/or CXXC. Thioredoxin reductase catalyzes the disulfide reduction of oxidized thioredoxin, using NADPH via the FAD molecule and the redox-active cysteine residues (Waksman et al., 1994). Initially thioredoxin reductases were divided into two sub-groups (low and high molecular weight TrxR) based on the absence or presence of a dimerization domain (Williams et al., 2000). However, there are some thioredoxin reductase enzymes with distinct extra domains that do not fit well into these two classes. Therefore, based on structural and biochemical considerations, we proposed that thioredoxin reductases should be divided into five sub-classes. In spite all these differences, all thioredoxin reductases share a common core, containing two domains (a NADPH-binding domain and a FAD-binding domain) and two redox centers: a FAD molecule and a dithiol-disulfide group (Oliveira et al., 2010).

Cytosolic thioredoxin system from yeast *Saccharomyces cerevisiae* is composed of one low molecular weight thioredoxin reductase (yTrxR1) and two thioredoxin enzymes (yTrx1 and yTrx2). Interestingly, most thioredoxin systems are composed of only one thioredoxin. yTrx1 and yTrx2 share 78% amino-acid identity and were initially considered as fully redundant enzymes. We are investigating whether these two oxido-reductases have specific roles. The expression of yTrx2 is highly inducible by peroxides in a process mediated by Yap1, whereas expression of yTrx1 is more constitutive (Kuge and Jones, 1994; Lee et al., 1999). The relevance of the cytosolic thioredoxin system from yeast can be attested to by the fact that deletion of ScTrxR1 gene renders yeast inviable (Giaever et al., 2002). Some of their targets include at least three peroxiredoxins (Tsa1, Tsa2 and Ahp1), methionine sulfoxide reductase, ribonucleotide reductase, a Cys-based peroxidase involved in the oxidative stress response (Gpx3-Orp1) (Fourquet et al., 2008) and the system involved in sulfate assimilation (PAPS). yTrx1 and yTrx2 present specificity towards their targets, *i.e.*, they cannot reduce all disulfide bonds. Some protein-protein interactions should occur in order for a protein disulfide reduction to take place. Once yTrx1 (or yTrx2) reduces a target disulfide bond, it gets oxidized. While studying the reduction of yTrx1 (or yTrx2) by yTrxR1, we observed that this flavoprotein exhibited remarkable specificity, *i.e.*, yTrxR1 only reduces yeast thioredoxins (yTrx1 and yTrx2), but not mammalian or bacterial thioredoxins (Oliveira et al., 2010). yTrxR1 can also reduce yeast mitochondrial thioredoxin (yTrx3). Probably this species-specificity phenomenon involves recognition of certain amino acid residues by yTrxR1 through protein-protein interactions.

The identification of protein–protein interactions is a major challenge in cell biology. The interactions for the various pathways are specific, directing signals to specific targets. Thiol-based systems are emerging as relevant pathways in signal transduction. Because there are several thiol-disulfide oxido-reductases in each genome, it is reasonable to think that each one of them interacts with different partners. Although all thioredoxin reductases catalyze the same overall reaction (*i.e.*, reduction of thioredoxin at the expense of NADPH), apparently the species-specificity phenomenon is restricted to the low molecular weight enzymes. This is probably because high molecular weight thioredoxin reductases have an external selenocysteine residue (a cysteine analog with a selenium instead of the sulfur atom

Site-Directed Mutagenesis as a Tool to Characterize Specificity in Thiol-Based Redox Interactions Between Proteins and Substrates

75

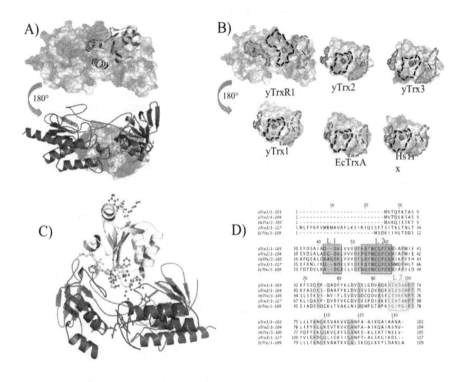

Fig. 1. **Structural features involved in the species–specificity phenomenon.** (A) Theoretical representation of the yTrx-yTrxR complex. The electrostatic surface of yTrxR1 is depicted at the top (Red = negatively charged atoms; blue = positively charged atoms; white = no charge) and yTrx1 is represented by the cartoon (yellow). At the bottom is shown the yTrxr1 electrostatic surface and yTrxr1 (cartoon-blue). (B) Comparison of electrostatic surfaces among five distinct thioredoxin enzymes, three of them from yeast. yTrx = thioredoxin from *S. cerevisiae*; EcTrxA = thioredoxin A from *E. coli*; and HsTrx = thioredoxin from *H. sapiens*. (C) Loop 2 (grape) and loop 3 (green) are in close proximity to thioredoxin reductase. (D) Amino-acid alignment among five thioredoxin enzymes. Three loops are candidates for physical interaction with thioredoxin reductase. Since the loop 3 (L3) amino-acid sequences display higher variability than the loop 2 (L2) sequences, loop 3 is implicated in the species–specificity phenomenon.

in the side chain) that can reduce target substrates with low physical interaction. Therefore, species specificity probably requires extensive protein-protein interactions. As mentioned above, thioredoxin reductase 1 from *S. cerevisiae* (yTrxR1 = yTrr1) can reduce cytosolic and mitochondrial thioredoxins from yeast; but it cannot reduce thioredoxin from *Escherichia coli* or from *Homo sapiens*.

Since thioredoxins present high sequence similarity, we are interested in identifying factors involved in species–specific interactions. As mentioned above, we obtained the structures of all proteins belonging to the cytosolic thioredoxin system from the yeast *S. cerevisiae* (yTrx1 = PDB 2I9H; yTrx2 = PDB 2HSY; yTrxR1 = PDB 3ITJ). This allowed us to test models for protein–protein interactions. The analysis indicated that complementary electrostatic surfaces between yTrxR1 and yTrxs are partially responsible for the species-specificity phenomenon (Fig. 1A). Furthermore, residues that belong to loop 3 appear to be directly related to protein-protein interactions (Fig. 1B). Indeed site-directed mutagenesis was a valuable tool for testing hypotheses raised by crystal structure analysis and by biochemical assays (Oliveira et al., 2010).

3.1.2 Aspects involved in the high oxido-reductase activity of yeast Glutaredoxin 2 in comparison with yeast Glutaredoxin 1

Like thioredoxin, glutaredoxin enzymes are thiol-disulfide oxido-reductases, whose genes are widespread among both eukaryotic and prokaryotic genomes. These small, heat stable enzymes are ubiquitously distributed and endowed with disulfide reductase activity (Discola et al., 2009). In the case of the yeast *S. cerevisiae*, eight isoforms have been identified so far. Three of them are dithiolic glutaredoxins, with two vicinal cysteines in a CXXC motif (mostly CPYC). The other five are monothiolic enzymes. They are characterized less well and will not be considered here.

yGrx1 and yGrx2 are the two major dithiolic glutaredoxins from *S. cerevisiae*. They display high amino acid sequence similarity to each other (85%). These enzymes can reduce disulfide bonds through two distinct mechanisms. In the most studied one, a mixed disulfide between a target protein and GSH is reduced by the monothiolic mechanism. In this case, only the N-terminal Cys of the CXXC motif takes part (Figure 2, reactions f and g). The most used assay to measure glutaredoxin activity, the HED (β-hydroxyethyl disulfide) assay, operates through this monothiolic pathway. Alternatively, a glutaredoxin with two Cys residues can reduce disulfides through the dithiolic pathway (Figure 2, reactions a–e).

The monothiolic pathway has received increased attention because it appears to control the levels of glutathionylated enzymes in cells. Glutathiolation is an emerging post-translational modification. This modication protects reactive Cys residues from irreversible oxidation to the sulfinic (RSO_2H) or sulfonic (RSO_3H) states. In analogy to phosphatases, glutaredoxins catalyze removal of the glutathionyl moiety and thereby regulate signaling processes (Gallogly and Mieyal, 2007). Like phosphorylation and other post-translational modifications, glutathionylation is reversible. To evaluate if a specific glutathiolation event is regulatory, some criteria were proposed, such as (1) a change in the activity of the target protein; (2) occurrence in response to a stimulus; or (3) occurrence at a physiological GSH/GSSG ratio (normally high in the cytosol), with both the modification and its disappearance being fast (Gallogly and Mieyal, 2007). Some proteins that fulfill these criteria are actin (Wang et al., 2001; Wang et al., 2003), Ras (Adachi et al., 2004) and Protein Tyrosine Phosphatase (Kanda et al., 2006).

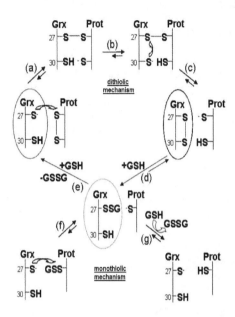

Fig. 2. **Mechanisms of disulfide reduction by glutaredoxin.** Dithiolic: Reactive Cys (in thiolate form) from glutaredoxin (Grx) performs a nucleophilic attack (S_N2 type) on a disulfide of a target proteins, leading to the formation of a mixed disulfide (reaction a); a thiolate is formed in the second Cys (reaction b) and this thiolate performs a nucleophilic attack (S_N2 type) on the mixed disulfide, generating a intramolecular disulfide on glutaredoxin (reaction c). Reduction of glutaredoxin takes places by two consecutive reactions with GSH (reactions d and f). Monothiolic: Reactive Cys (in thiolate form) from glutaredoxin (Grx) performs a nucleophilic attack (S_N2 type) on a mixed disulfide between GSH and a target protein, leading to the formation of glutathiolated glutaredoxin (reaction f); reduction takes place by reaction with a second GSH molecule (reaction g). Reaction g is considered the rate-limiting step in the monothiolic pathway (Srinivasan et al., 1997).

In collaboration with Dr. Demasi, our group has shown that the proteasome is also post-translationally modified by glutathiolation in response to oxidative stress (Demasi et al., 2001; Demasi et al., 2003) and also that glutaredoxin can reduce the mixed disulfide bond between the proteasome and GSH (Silva et al., 2008). Site-directed mutagenesis of Cys residues in the 20S proteasome is underway in order to clarify mechanistic details of this process. Therefore, it is relevant to comprehend features that control the deglutathionylase activity of glutaredoxins (Figure 2, reactions f and g) to better appreciate the function of this post-translational modification in cell biology.

In this regard, it was relevant to observe that the two main dithiolic glutaredoxins from yeast display markedly distinct monothiolic (HED assay) specific activities (Discola et al., 2009). Although these two enzymes share a high degree of similarity in their amino acid sequences, yGrx2 is two orders of magnitude more active than yGrx1 (Discola et al, 2009). These data are consistent with results from studies with knockout strains (*i.e.*, strains with null alleles) that indicate that yGrx2 accounts for most of the oxido-reductase activity observed in yeast extracts (Luikenhuis et al., 1998). In an attempt to gain insights on this phenomenon, our group obtained two crystallographic structures of yGrx2 (intramolecular disulfide = PDB 3D4M; mixed disulfide with GSH = PDB 3D5J, both of which are related to the short form yGrx2) and compared them with the crystal structures of yGrx1 (reduced = PDB 2JAD; mixed disulfide with GSH = PDB 2JAC) available in the literature (Håkansson and Winther, 2007). The overall structures are highly similar (Fig. 3A). However, differences in the active sites were hypothesized to be involved in the distinct catalytical efficiencies between yGrx1 and yGRx2 (Fig. 3B). In order to obtain the structures of these complexes, it was necessary to mutate the C-terminal Cys (Cys30) to Ser in order to slow reaction d (reverse) (see d in Fig. 2). The analysis of the structures of yGrx1C30S and yGrx2C30S (short isoform) in complex with GSH revealed that the distances between Ser30 (Cys 30 in both yGrx1 and yGrx2, short isoform, wild-type proteins) and the reactive Cys (Cys47) are markedly distinct (3.47 Å in yGrx1C30S and 5.14 Å in yGrx2C30S).

A)   B)

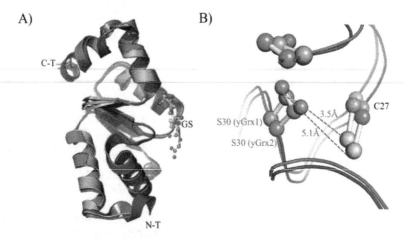

Fig. 3. **Crystal structures of dithiolic glutaredoxins from yeast.** (A) Cartoon representation of overall glutaredoxin structures. (Red= yGrx1C30S, PDB code 2JAC; Green = yGrx2 disulfide, PDB code 3D4M; Blue = yGrx2C30S, PDB code = 3D5J); (B) Active sites of the complex with glutathione. The distances between the resolving Cys and the reactive cysteine (sulfur atoms colored in yellow) are shown by dashed lines. Colors are defined in A.

In principle, any factor that would slow reaction d (reverse) (see d in Fig.2) should favor reaction g and, consequently, the monothiolic activity, which is the mechanism by which the HED assay operates. According to this hypothesis, anything that increases the distance between the two sulfur atoms of the CXXC motif should favor the monothiolic activity and, consequently, the rates in the HED assay. Accordingly, Ser23 is in close proximity to Ser30 in yGRx2 (short isoform). This interaction between two serine residues is probably stabilizing a configuration, in which the distances between the two sulfur atoms would be high in the wild-type yGrx2, thereby accounting for its high catalytical efficiency (Discola et al., 2009). In contrast, yGrx1 has an Ala residue at position 23; and this side chain cannot make a salt bridge. In this case, the distances for residues 27 and 30 (cysteines in the wild-type protein) are short. A short distance between the sulfur atoms favors the dithiolic mechanism over the monothiolic mechanism (Fig. 2). This hypothesis was tested by site-directed mutagenesis. Indeed the relevance of a serine at position 23 for the monothiolic activity was demonstrated (Discola et al., 2009).

Probably biochemical and structural features other than a serine/alanine residue at position 23 are involved with the higher catalytical efficiency of yGrx2 over yGrx1. Indeed Ser89 in yGrx2 (short isoform) and Asp89 in yGrx1 were recently implicated in the different catalytical properties of these two oxido-reductases (Li et al., 2010). Ser89 is involved in the binding of GSH in glutaredoxin (Discola et al., 2009). The authors also employed site-directed mutagenesis to show that their hypothesis was correct (Li et al., 2010).

Since glutathionylation is emerging as a key concept in redox signaling, it is reasonable that the combined approach of biochemical and structural assays together with site-directed mutagenesis will be followed to establish the involvement of other factors in this post-translational modification.

3.1.3 Site-directed mutagenesis to characterize residues that allow reduction of 1-Cys peroxiredoxin by ascorbate

Peroxiredoxins are ubiquitous, Cys-based peroxidases, whose importance is underlined by their high abundance and their involvement in multiple cellular processes probably related to their capacity to decompose hydroperoxides (Rhee and Woo, 2011; Wood et al., 2003). Indeed several groups have shown independently that peroxiredoxins compete with heme-peroxidases and Se-GSH peroxidases for hydroperoxides (Horta et al., 2010; Ogosucu et al., 2007; Parsonage et al., 2005; Toledo Jr et al., 2011). As a consequence of their high abundance and reactivity, peroxiredoxins are major sinks for peroxides (Winterbourn and Hampton, 2008).

A peroxiredoxin can be classified as a 1-Cys or 2-Cys Prx, depending on the number of Cys residues that participate in the catalytic cycle (Rhee and Woo, 2011; Wood et al., 2003). For most 2-Cys Prxs, the biological reductant is thioredoxin. For 1-Cys Prxs, the situation is far more complex. In many cases the identity of the reductant is not known. Our group has shown that ascorbate can support the peroxidase activity of 1-Cys Prx. This represented a change of the thiol-specific antioxidant paradigm (Monteiro et al., 2007).

Since 1-Cys and 2-Cys Prxs share amino acid sequence similarity, we asked ourselves which amino acids are responsible for the ability of 1-Cys Prx enzymes to accept ascorbate as the electron donor. Through a multiple approach involving amino acid sequence alignment, mass spectrometry and enzymatic assays, we postulated that two features are required: (1) the absence of a Cys involved in disulfide formation (resolving Cys) and (2) the presence of a His residue fully conserved in 1-Cys Prxs and absent in the 2-Cys Prx counterparts. By site-directed mutagenesis, we were able to engineer a 2-Cys Prx to be reducible by ascorbate by taking into account the two factors described above (Monteiro et al., 2007). Further studies are underway in order to comprehend the physiological significance of ascorbating acting as a reducing agent for 1-Cys Prx.

3.2 Cys-based proteins from bacteria

3.2.1 Residues of Peroxiredoxin Q involved in redox-dependent secondary structure change

Our group is also interested in the analyses of Cys-based proteins from bacteria, as a consequence of our participation in the genome-sequencing project of the bacterium *Xylella fastidiosa*. *X. fastidiosa* is a gram-negative bacterium that is the etiologic agent of several plant diseases, such as Citrus Variegated Chlorosis, which imposes great losses in orange production in Brazil (Lambais, 2000). *X. fastidiosa* also causes Pierce disease in grapevines, phony peach disease, and leaf scorch diseases in almond and oleander (Hendson et al., 2001).

Animal and plant hosts generate oxidative insults against pathogens, such as *X. fastidiosa*, in an attempt to avoid infection. The oxidants include hydrogen peroxide, organic hydroperoxides, and peroxynitrite (Koszelak-Rosenblum et al., 2008; Tenhaken et al., 1999; Wrzaczek et al., 2009). To counteract this host response, bacteria present a large repertoire of antioxidants, including Cys-based peroxidases (Horta et al., 2010). Therefore, in principle, any intervention that results in the decrease of antioxidants from pathogens can have a therapeutic property. Indeed the mechanism of action of several antibiotics is based on the generation of oxidants (Kohanski et al., 2010).

After completion of the genome-sequencing project of *X. fastidiosa* (Simpson et al., 2000), we decided to characterize Cys-based peroxidases from this plant pathogen. Analysis of the *X. fastidiosa* genome revealed the presence of five genes that encode proteins potentially involved in hydroperoxide decomposition: one catalase, one glutathione peroxidase (GPx), one organic hydroperoxide resistance protein (Ohr) and two peroxiredoxins (AhpC and PrxQ), both of which probably display the 2-Cys Prx mechanism (Horta et al., 2010). All of them, except GPx protein, were identified in the whole-cell extract and extracellular fraction of the citrus-isolated strain 9a5c (Smolka et al., 2003). We decided to characterize peroxiredoxins from *X. fastidiosa*.

As noted above (Section 3.1.3), peroxiredoxins can be classified into two groups, 2-Cys Prxs and 1-Cys Prxs, depending on the mechanism of catalysis. Besides this mechanistic classification, others were proposed that are based on amino acid sequence similarity. Later structural features were incorporated into the classification proposals. They provided insights on the evolution of proteins within the Trx superfamily, which includes the Prxs (Copley et al., 2004; Nelson et al., 2011). Adopting the classification described in Copley et al. (2004) for Prx classes, class 1 is the most ancestral, but the least characterized of all 4 classes. The other 3 classes of Prxs were derived from those of class 1. We therefore decided to investigate a class 1 Prx, peroxiredoxin Q (XfPrxQ) from *X. fastidiosa* (Horta et al., 2010).

Historically all classes of Prx have been considered only moderately reactive. The reason was that their catalytic efficiencies (k_{cat}/Km) toward hydroperoxides, as determined by steady-state kinetics, were in the 10^4-10^5 M^{-1} s^{-1} range. In contrast, selenocysteine-containing GPx (10^8 M^{-1} s^{-1}) and heme-containing catalases (10^6 M^{-1} s^{-1}) presented considerably higher values (Wood et al., 2003). More recently, with the development of new assays, Prx enzymes were considered as reactive as selenium- and heme-containing proteins (Ogusucu et al., 2007; Parsonage et al., 2005; Trujillo et al., 2007). At that time, only class 3 Prx enzymes and class 4 Prx enzymes (composed mostly of typical 2-Cys Prx enzymes, but also some 1-Cys Prx proteins) were analyzed by these assays. Consequently, the catalytic efficiencies for enzymes of the other Prx classes remained to be determined. Through a competitive-kinetics approach (Toledo et al., 2011), we demonstrated that the second-order rate constants of the peroxidase reactions of XfPrxQ with hydrogen peroxide and peroxynitrite lay in the order of 10^7 and 10^6 M^{-1} s^{-1}, respectively. These reactions are as fast as the most efficient peroxidases. Furthermore, the catalytic cycle of XfPrxQ was elucidated by multiple approaches, such as X-ray crystallography, circular dichroism, biochemical assays, mass spectrometry, and site-directed mutagenesis. Using data obtained by site-directed mutagenesis, we were able to propose a model for the redox-dependent structural changes in PrxQ proteins (Fig. 4) that was consistent with all of our data and data from the literature (Horta et al., 2010). Site-

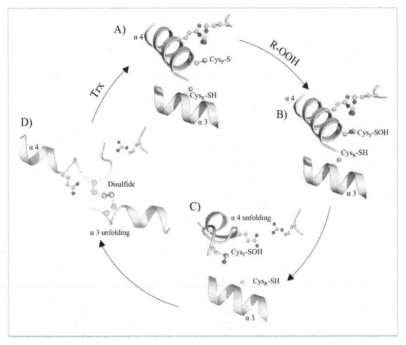

Fig. 4. **Model for the redox-dependent conformational changes in XfPrxQ.** Proposed sequence of structure snapshots along the catalytic cycle of the PrxQ subfamily proteins. The protein is represented in cartoon (light green) and residues are represented as sticks. Atoms are colored as follow: C=green, O=red, N=blue and S=orange. Peroxidatic and resolving cysteines are indicated as CysP and CysR, respectively. (A) Reduced species based on the crystal structure of XfPrxQ C47S. (B) and (C) Hypothetical conformational intermediates based on circular dichroism data (Horta et al., 2010). (D) Oxidized species based on the crystal structure of Bcp from *X. campestris* (PDB code = 3GKK).

directed mutagenesis revealed that Cys47 is the center responsible for the changes in secondary structure measured by circular dichroism (Horta et al., 2010).

3.2.2 Site-directed mutagenesis as a valuable tool to characterize a new antioxidant protein: Organic Hydroperoxide Resistance Protein

During the annotation of the X. *fastidiosa* genome (Simpson et al., 2000), the function of one gene caught our attention: *ohr*. What was reported at that time was that the deletion of *ohr* gene in X. *campestris pv. Phaseoli* rendered those cells sensitive to the oxidative insult by organic hydroperoxides, but not by hydrogen peroxide (Mongkolsuk et al., 1998). Furthermore, the transcription of the *ohr* gene was specifically induced by organic hydroperoxides, such as *tert*-butyl-hydroperoxide (Mongkolsuk et al., 1998). Therefore, this gene was named *ohr* (organic hydroperoxide resistance gene). However, the biological activity for the corresponding protein was not known.

Alignment of the deduced amino acid sequences of putative Ohr proteins from several bacteria revealed the presence of two fully conserved Cys residues. Therefore, we hypothesized that the *ohr* gene probably encodes a Cys-based, thiol-dependent peroxidase. In order to test this hypothesis, recombinant Ohr was obtained by cloning and expressing the *ohr* gene from X. *fastidiosa*. Indeed it displayed a thiol-dependent peroxidase (Cussiol et al., 2003). Remarkably, the peroxidase activity of Ohr was specifically supported by dithiols, such as DTT, but not by monothiols, such as 2-mercapthoethanol. In contrast, both mono- and dithiols support the enzymatic activity of peroxiredoxins and glutathione peroxidase. Furthermore, Ohr shows high activity towards organic hydroperoxides. Site-directed mutagenesis of the two conserved Cys residues unequivocally revealed that the most N-terminal one (Cys62) is the redox center (Cussiol et al., 2003). Another major achievement was the elucidation of the X-ray structure of Ohr (Oliveira et al., 2006). It showed a unique structure. The "Ohr fold" was quite different from the thioredoxin fold that is present in peroxiredoxin and the GSH peroxidases. At the same time, the structure and biochemical activity of the Ohr from *Pseudomonas aeruginosa* was elucidated (Lesniak et al., 2002). Essentially it has the same features described above.

In contrast to peroxiredoxins and GSH peroxidases, in which the reactive Cys residue is solvent-exposed, the reactive Cys residue in Ohr is buried in the polypeptide chain. The microenvironment where the reactive Cys is located is surrounded by several hydrophobic residues that probably confer to Ohr higher affinity for organic hydroperoxides. Proteins with folds similar to Ohr are only present in bacteria. This fact might indicate that this peroxidase is a target for drug development. Another unique property of Ohr is that only lipoamide, and neither GSH nor thioredoxin can support its Cys-based peroxidase activity (Cussiol et al., 2010). Due to these distinct properties, we are searching for Ohr inhibitors. Therefore, we are pursuing the characterization of enzyme-substrate interactions. We found in the active site of Ohr an electronic density of the polyethylene glycol molecule. Polyethylene glycol is a polymeric compound with elongated shape that was used in the crystallization trials. Since peroxides derived from fatty acids present an elongated structure and fit very well into this electronic density, we proposed that this kind of substrate may be the physiological target of the Ohr enzyme. Amino acid residues possibly involved in enzyme–substrate interactions were identified (Oliveira et al., 2006). Currently site-directed mutagenesis of these residues is underway in order to test the hypothesis that they are involved in the binding of organic hydroperoxides. This information will be relevant in the search for Ohr inhibitors. We have already found some chemicals that can inhibit Ohr and also inhibit the growth of of X. *fastidiosa*.

4. Concluding remarks

In the characterization of Cys-based proteins, the involvement of amino acids in catalysis was analyzed by several enzymatic and biochemical assays, as well as by X-ray crystallography. Using an approach that combined these methodologies with site-directed mutagenesis allowed several hypotheses to be raised. Research is currently testing them.

5. References

Adachi T, Pimentel DR, Heibeck T, Hou X, Lee YJ, Jiang B, Ido Y, Cohen RA (2004) S-Glutathiolation of Ras mediates redox-sensitive signaling by angiotensin II in vascular smooth muscle cells. *J Biol Chem*. 279, 29857–29862.

Amorim GC, Pinheiro AS, Netto LE, Valente AP, Almeida FC (2007) NMR solution structure of the reduced form of thioredoxin 2 from *Saccharomyces cerevisiae*. *J Biomol NMR*. 38, 99–104.

Benesch RE, Benesch R (1955) The acid strength of the -SH group in cysteine and related compounds. *J Amer Chem Soc*. 77, 5877–5881.

Berndt C, Lillig CH, Holmgren A (2007) Thiol-based mechanisms of the thioredoxin and glutaredoxin systems: implications for diseases in the cardiovascular system. *Am J Physiol Heart Circ Physiol*. 292, H1227–1236.

Copley SD, Novak WR, Babbitt PC (2004) Divergence of function in the thioredoxin fold suprafamily: evidence for evolution of peroxiredoxins from a thioredoxin-like ancestor. *Biochemistry*, 43, 13981–13995.

Cussiol JR, Alegria TG, Szweda LI, Netto LE (2010) Ohr (organic hydroperoxide resistance protein) possesses a previously undescribed activity, lipoyl-dependent peroxidase. *J Biol Chem*. 285, 21943–21950.

Cussiol JR, Alves SV, Oliveira MA, Netto LE (2003) Organic hydroperoxide resistance gene encodes a thiol-dependent peroxidase. *J Biol Chem*. 278, 11570–11578.

Demasi M, Shringarpure R, Davies KJ (2001) Glutathiolation of the proteasome is enhanced by proteolytic inhibitors. *Arch Biochem Biophys*. 389, 254–263.

Demasi M, Silva GM, Netto LE (2003) 20 S proteasome from *Saccharomyces cerevisiae* is responsive to redox modifications and is S-glutathionylated. *J Biol Chem*. 278, 679–685.

Discola KF, Oliveira MA, Rosa Cussiol JR, Monteiro G, Bárcena JA, Porras P, Padilla CA, Guimarães BG, Netto LE (2009) Structural aspects of the distinct biochemical properties of glutaredoxin 1 and glutaredoxin 2 from *Saccharomyces cerevisiae*. *J Mol Biol*. 385, 889–901.

Ferrer-Sueta G, Manta B, Botti H, Radi R, Trujillo M, Denicola A (2011) Factors affecting protein thiol reactivity and specificity in peroxide reduction. *Chem Res Toxicol*. 24, 434–450.

Gallogly MM, Mieyal JJ (2007) Mechanisms of reversible protein glutathionylation in redox signaling and oxidative stress. *Curr Opin Pharmacol*. 7, 381–391.

Giaever G, Chu AM, Ni L, Connelly C, Riles L, Véronneau S, Dow S, Lucau-Danila A, Anderson K, André B, Arkin AP, Astromoff A, El-Bakkoury M, Bangham R, Benito

R, Brachat S, Campanaro S, Curtiss M, Davis K, Deutschbauer A, Entian KD, Flaherty P, Foury F, Garfinkel DJ, Gerstein M, Gotte D, Güldener U, Hegemann JH, Hempel S, Herman Z, Jaramillo DF, Kelly DE, Kelly SL, Kötter P, LaBonte D, Lamb DC, Lan N, Liang H, Liao H, Liu L, Luo C, Lussier M, Mao R, Menard P, Ooi SL, Revuelta JL, Roberts CJ, Rose M, Ross-Macdonald P, Scherens B, Schimmack G, Shafer B, Shoemaker DD, Sookhai-Mahadeo S, Storms RK, Strathern JN, Valle G, Voet M, Volckaert G, Wang CY, Ward TR, Wilhelmy J, Winzeler EA, Yang Y, Yen G, Youngman E, Yu K, Bussey H, Boeke JD, Snyder M, Philippsen P, Davis RW, Johnston M (2002) Functional profiling of the *Saccharomyces cerevisiae* genome. *Nature*, 418, 387–391.

Håkansson KO, Winther JR (2007) Structure of glutaredoxin Grx1p C30S mutant from yeast. *Acta Crystallog Sect D: Biol Crystallogr*. 63, 288–294.

Hall A, Parsonage D, Poole LB, Karplus PA (2010) Structural evidence that peroxiredoxin catalytic power is based on transition-state stabilization. *J Mol Biol*. 402, 194–209.

Hayashi T, Ueno Y, Okamoto T (1993) Oxidoreductive regulation of nuclear factor kappa B. Involvement of a cellular reducing catalyst thioredoxin. *J Biol Chem*. 268, 11380–11388.

Hendson M, Purcell AH, Chen D, Smart C, Guilhabert M, Kirkpatrick B (2001) Genetic diversity of Pierce's disease strains and other pathotypes of *Xylella fastidiosa*. *Appl Environ Microbiol*. 67, 895–903.

Horta BB, de Oliveira MA, Discola KF, Cussiol JR, Netto LE (2010) Structural and biochemical characterization of peroxiredoxin Qbeta from *Xylella fastidiosa*: catalytic mechanism and high reactivity. *J Biol Chem*. 285, 16051–16065

Jacob C, Giles GI, Giles NM, Sies H (2003) Sulfur and selenium: the role of oxidation state in protein structure and function. *Angew Chem Int Ed Engl*. 42, 4742–4758.

Jones DP (2006) Redefining oxidative stress. *Antioxid Redox Signal*. 8, 1865–1879.

Kanda M, Ihara Y, Murata H, Urata Y, Kono T, Yodoi J, Seto S, Yano K, Kondo T (2006) Glutaredoxin modulates platelet-derived growth factor-dependent cell signaling by regulating the redox status of low molecular weight protein-tyrosine phosphatase. *J Biol Chem*. 281, 28518–28528.

Kohanski MA, DePristo MA, Collins JJ.(2010) Sublethal antibiotic treatment leads to multidrug resistance via radical-induced mutagenesis. *Mol Cell*. 12;311-320.

Koszelak-Rosenblum M, Krol AC, Simmons DM, Goulah CC, Wroblewski L, Malkowski MG (2008) *J Biol Chem*. 283, 24962–24971.

Kuge S, Jones N (1994) YAP1 dependent activation of TRX2 is essential for the response of *Saccharomyces cerevisiae* to oxidative stress by hydroperoxides. *EMBO J*. 13, 655–664.

Lambais MR, Goldman MH, Camargo LE, Goldman GH (2000) A genomic approach to the understanding of *Xylella fastidiosa* pathogenicity. *Curr Opin Microbiol*. 3, 459–462.

Lee J, Godon C, Lagniel G, Spector D, Garin J, Labarre J, Toledano MB (1999) Yap1 and Skn7 control two specialized oxidative stress response regulons in yeast. *J Biol Chem*. 274, 16040–16046.

Lesniak J, Barton WA, Nikolov DB (2002) Structural and functional characterization of the *Pseudomonas* hydroperoxide resistance protein Ohr. *EMBO J*. 24, 6649–6659.

Li WF, Yu J, Ma XX, Teng YB, Luo M, Tang YJ, Zhou CZ (2010) Structural basis for the different activities of yeast Grx1 and Grx2. *Biochim Biophys Acta*. 1804, 1542–1547.

Lindahl M, Mata-Cabana A, Kieselbach T (2011) The disulfide proteome and other reactive cysteine proteomes: analysis and functional significance. *Antioxid Redox Signal*. 12, 2581–2642.

Luikenhuis S, Perrone G, Dawes IW, Grant CM (1998) The yeast *Saccharomyces cerevisiae* contains two glutaredoxin genes that are required for protection against reactive oxygen species. *Mol Biol Cell*, 9, 1081–1091.

Marino SM, Gladyshev VN (2011) Redox biology: computational approaches to the investigation of functional cysteine residues. *Antioxid Redox Signal*. 15, 135–146.

Matthews JR, Wakasugi N, Virelizier JL, Yodoi J, Hay RT (1992) Thioredoxin regulates the DNA binding activity of NF-kappa B by reduction of a disulphide bond involving cysteine 62. *Nucleic Acids Res*. 20, 3821–3830.

Monteiro G, Horta BB, Pimenta DC, Augusto O, Netto LE (2007) Reduction of 1-Cys peroxiredoxins by ascorbate changes the thiol-specific antioxidant paradigm, revealing another function of vitamin C. *Proc Natl Acad Sci USA*. 104, 4886–4891.

Nagy P, Karton A, Betz A, Peskin AV, Pace P, O'Reilly RJ, Hampton MB, Radom L, Winterbourn CC (2011) Model for the exceptional reactivity of peroxiredoxins 2 and 3 with hydrogen peroxide: a kinetic and computational study. *J Biol Chem*. 286, 18048–18055.

Nelson KJ, Knutson ST, Soito L, Klomsiri C, Poole LB, Fetrow JS (2011) Analysis of the peroxiredoxin family: using active-site structure and sequence information for global classification and residue analysis. *Proteins*, 79, 947–964.

Netto LE, Oliveira MA, Monteiro G, Demasi AP, Cussiol JR, Discola KF, Demasi M, Silva GM, Alves SV, Faria VG, Horta BB (2007) Reactive cysteine in proteins: protein folding, antioxidant defense, redox signaling and more. *Comp Biochem Physiol C Toxicol Pharmacol*. 146, 180–193.

Ogusucu R, Rettori D, Munhoz DC, Netto LE, Augusto O (2007) Reactions of yeast thioredoxin peroxidases I and II with hydrogen peroxide and peroxynitrite: rate constants by competitive kinetics. *Free Radic Biol Med*. 42, 326–334.

Oliveira MA, Discola KF, Alves SV, Medrano FJ, Guimarães BG, Netto LE (2010) Insights into the specificity of thioredoxin reductase-thioredoxin interactions. A structural and functional investigation of the yeast thioredoxin system. *Biochemistry*, 49, 3317–3326.

Oliveira MA, Guimarães BG, Cussiol JR, Medrano FJ, Gozzo FC, Netto LE (2006) Structural insights into enzyme-substrate interaction and characterization of enzymatic intermediates of organic hydroperoxide resistance protein from *Xylella fastidiosa*. *J Mol Biol*. 359, 433–445.

O'Neill JS, Reddy AB (2011) Circadian clocks in human red blood cells. *Nature*, 469, 498–503.

Parsonage D, Youngblood DS, Sarma GN, Wood ZA, Karplus PA, Poole LB (2005) Analysis of the link between enzymatic activity and oligomeric state in AhpC, a bacterial peroxiredoxin. *Biochemistry*, 44, 10583–10592.

Peskin AV, Low FM, Paton LN, Maghzal GJ, Hampton MB, Winterbourn CC (2007) The high reactivity of peroxiredoxin 2 with H_2O_2 is not reflected in its reaction with other oxidants and thiol reagents. *J Biol Chem*. 282, 11885–11892.

Pinheiro AS, Amorim GC, Netto LE, Almeida FC, Valente AP (2008) NMR solution structure of the reduced form of thioredoxin 1 from *Sacharomyces cerevisiae*. *Proteins*, 70, 584–587.

Powis G, Montfort WR (2001) Properties and biological activities of thioredoxins. *Annu Rev Biophys Biomol Struct*. 30, 421–455.

Rhee SG, Woo HA (2011) Multiple functions of peroxiredoxins: peroxidases, sensors and regulators of the intracellular messenger H_2O_2, and protein chaperones. *Antioxid Redox Signal*. 15, 781–794.

Saitoh M, Nishitoh H, Fujii M, Takeda K, Tobiume K, Sawada Y, Kawabata M, Miyazono K, Ichijo H (1998) Mammalian thioredoxin is a direct inhibitor of apoptosis signal-regulating kinase (ASK) 1. *EMBO J.* 17, 2596–2606.

Silva GM, Netto LE, Discola KF, Piassa-Filho GM, Pimenta DC, Bárcena JA, Demasi M (2008) Role of glutaredoxin 2 and cytosolic thioredoxins in cysteinyl-based redox modification of the 20S proteasome. *FEBS J.* 275, 2942–2955.

Simpson AJ et al (2000) The genome sequence of the plant pathogen *Xylella fastidiosa*. The *Xylella fastidiosa* Consortium of the Organization for Nucleotide Sequencing and Analysis. *Nature*, 406, 151–159.

Smolka MB, Martins-de-Souza D, Martins D, Winck FV, Santoro CE, Castellari, RR, Ferrari F, Brum IJ, Galembeck E, Della Coletta Filho H, Machado MA, Marangoni S, Novello JC (2003) Proteome analysis of the plant pathogen *Xylella fastidiosa* reveals major cellular and extracellular proteins and a peculiar codon bias distribution. *Proteomics*, 3, 224–237.

Srinivasan U, Mieyal PA, Mieyal, JJ (1997) pH profiles indicative of rate-limiting nucleophilic displacement in thioltransferase catalysis. *Biochemistry*, 36, 3199–3206.

Tenhaken R, Levine A, Brisson LF, Dixon RA, Lamb C (1995) Function of the oxidative burst in hypersensitive disease resistance. *Proc Natl Acad Sci USA.* 92, 4158–4163.

Toledo JC Jr, Audi R, Ogusucu R, Monteiro G, Netto LE, Augusto O (2011) Horseradish peroxidase compound I as a tool to investigate reactive protein-cysteine residues: from quantification to kinetics. *Free Radic Biol Med.* 50, 1032–1038.

Trujillo M, Ferrer-Sueta G, Thomson L, Flohé L, Radi R (2007) Kinetics of peroxiredoxins and their role in the decomposition of peroxynitrite. *Subcell Biochem.* 44, 83–113.

Waksman G, Krishna TSR, Williams CH Jr, Kuriyan J (1994) Crystal structure of *Escherichia coli* thioredoxin reductase refined at 2 Å resolution. Implications for a large conformational change during catalysis. *J Mol Biol.* 236, 800–816.

Wang J, Boja ES, Tan W, Tekle E, Fales HM, English S, Mieyal JJ, Chock PB (2001) Reversible glutathionylation regulates actin polymerization in A431 cells. *J Biol Chem.* 276, 47763–47766.

Wang J, Tekle E, Oubrahim H, Mieyal JJ, Stadtman ER, Chock PB (2003) Stable and controllable RNA interference: investigating the physiological function of glutathionylated actin. *Proc Natl Acad Sci USA.* 100, 5103–5106.

Williams CH Jr, Arscott LD, Muller S, Lennon BW, Ludwig ML, Wang PF, Veine DM, Becker K, Schirmer RH (2000) Thioredoxin reductase: Two modes of catalysis have evolved. *Eur J Biochem.* 267, 6110–6117.

Winterbourn CC, Metodiewa D (1999) Reactivity of biologically important thiol compounds with superoxide and hydrogen peroxide. *Free Radic Biol Med.* 27, 322–328.

Winterbourn CC, Hampton MB (2008) Thiol chemistry and specificity in redox signaling. *Free Radic Biol Med.* 45, 549–561.

Wood ZA, Schröder E, Robin Harris J, Poole LB (2003) Structure, mechanism and regulation of peroxiredoxins. *Trends Biochem Sci.* 1, 32–40.

Wrzaczek M, Brosché M, Kollist H, Kangasjärvi J (2009) Arabidopsis GRI is involved in the regulation of cell death induced by extracellular ROS. *Proc Natl Acad Sci USA.* 106, 5412–5417.

5

Directed Mutagenesis of Nicotinic Receptors to Investigate Receptor Function

Jürgen Ludwig, Holger Rabe, Anja Höffle-Maas,
Marek Samochocki, Alfred Maelicke and Titus Kaletta
Galantos Pharma GmbH
Germany

1. Introduction

Nicotinic acetylcholine receptors (nAChR) are the archetypes of drug receptors. In 1905, John Newport Langley introduced the concept of a receptive substance on the surface of skeletal muscle that mediated the action of a drug, such as nicotine and curare (a neurotoxin made from a plant) (Langley et al., 1905). He also proposed that these receptive substances were different in different species and tissues, and they undergo conformational changes in response to the respective drug. Today nAChRs are considered prototypes of receptors that function as integral signal transducers. They have the response element and the ion channel domain within the same molecular entity as the ligand-binding domain that is activated by acetylcholine. In contrast, muscarinic acetylcholine receptors (mAChR) are prototypic G protein-coupled receptors. They also sense molecules outside the cell, but they require the G proteins to induce cellular responses by coupling to intracellular signalling pathways. More important than their historical role, nicotinic receptors continue to be at the forefront of science, as they are drug targets for muscle and nerve diseases, such as Alzheimer's and Parkinson's diseases, Schizophrenia and Myasthenia gravis. Therefore, this receptor family serves as an excellent example for demonstrating the suitability of site-directed mutagenesis for investigating receptor function and exploring drug action.

The neurotransmitter acetylcholine binds to the extracellular domain of the nAChR and, consequently, opens the receptor-integral membrane channel for Na^+, K^+ and Ca^{2+} ions. The channel can close in two ways. (1) The acetylcholine dissociates from its extracellular binding site, a process that is enhanced by rapid cleavage of the neurotransmitter by the acetylcholine esterase in the synaptic cleft. Thus released, the neurotransmitter has only a short time to act on nAChRs before the signal is terminated again. (2) The channel closes spontaneously despite the presence of transmitter, a process called desensitization. Desensitization is a protective mechanism against exposure to acetylcholine and its agonists that is too long or too strong. Desensitization thus avoids excessive influx of ions into the cell, which can result in impairment of cellular function and cell death.

Ligand binding to nAChR usually occurs in the submillisecond range. The receptor-integral channel is opened only for a few milliseconds and is then closed. Nicotinic receptors are therefore extremely fast and efficient signal transducers. They play key roles in such life-important properties as muscle contraction and brain action.

1.1 Structure

A nAChR is composed of five subunits that form a central pore. The extracellular domain contains the binding site for the neurotransmitter acetylcholine (ACh). The four transmembrane helices from each subunit make up the integral ion channel. Depicted are some of the features described in this section. The C-loop is an important part of the acetylcholine binding site. The binding site of the modulating substance galantamine (GAL) is situated at another part of the C-loop. A phosphorylation site (P) is important for modulating the activity of the receptor. A glycosylation site (G) seems to play a role in cobratoxin resistance. Ivermectin and PNU-120596 are other substances that modulate the activity of the receptor, but they have different locations (IVE and PNU). The cytoplasmatic loop is important for receptor targeting.

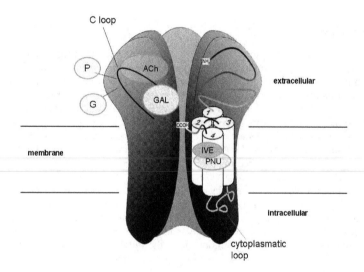

Fig. 1. Schematic representation of a nicotinic Acetylcholine Receptor (nAChR)

The nAChR is assembled from five subunits that are organized around a central pore. Seventeen homologous subunits ($\alpha 1$ – $\alpha 10$, $\beta 1$ – $\beta 4$, γ, δ and ε) are known in vertebrates (Zouridakis et al., 2009). This pool of subunits accounts for the vast number of nACh receptor subtypes that exhibit extensive functional diversity with respect to their pharmacological profile, spatiotemporal expression patterns and kinetic properties.

All subunits share the same architectural blueprint, *i.e.*, they consist of a large aminoterminal extracellular domain containing the name-giving cysteine loop, followed by a transmembrane domain and a small intracellular domain. The transmembrane domain consists of four transmembrane regions (TM1 to TM4). TM1, TM2, and TM3 are linked by two short loops. A long and highly variable intracellular loop occurs between TM3 and TM4. Except for those of the long intracellular loop, the amino acid residues are substantially conserved. This pool of subunits accounts for the vast number of nACh receptor subtypes that exhibit extensive functional diversity with respect to their pharmacological profile, spatiotemporal expression patterns and kinetic properties.

1.2 Use of directed mutagenesis in nicotinic acetylcholine receptor research

Site-directed mutagenesis is a powerful tool to investigate the role of individual amino acids within a protein, to understand the function of a given protein or to understand pharmacological interactions between the protein and compounds. However, modifying individual amino acids or larger parts of a protein with unknown function bears a number of risks. (1) The impact of a particular mutation might appear to be subtle, but it could also lead to a dysfunctional receptor or to the failure of receptor assembly. (2) A mutation might exert an effect on a distant site of the molecule due to conformational changes. This might be interpreted wrongly. (3) The mutated protein might adopt new properties that are unrelated to the natural protein. Hence, it is advisable always to design a battery of receptor mutants or chimeras. It is also important to include similar mutants, either by mutating an amino acid to several different amino acids or by creating similar chimeras. If the results of these mutants are consistent, the risk of erroneously assigning a wrong function can be reduced. Combining mutagenesis studies with other approaches, such as molecular modelling, will further substantiate a hypothesis. A number of such cases will be described in Section 2. Here, the two major mutant types will be described.

1.2.1 Single amino acid changes

Single amino acid changes can be used to investigate the role of amino acids in the binding of the natural ligands and drugs. Furthermore, this type of mutagenesis can be applied to cases in which computer algorithms have predicted motifs, such as glycosylation or phosphorylation sites. Other possible applications are assembly, the targeting of the receptor, or the identification of signals for expression.

A common way to designate a mutant is to use a letter-number-letter scheme. The first letter indicates the wild-type amino acid, using the one-letter code (see Table 1 in the chapter by Figurski et al. for the amino acid codes); the number refers to the position of the amino acid in the protein; and the final letter designates the amino acid that now occupies the position of the original amino acid (*e.g.*, T197A refers to a mutant in which threonine at position 197 was changed to alanine).

1.2.2 Chimeric receptors

Chimeric receptors combine parts of different receptors. This type of mutagenesis is useful in cases in which, instead of a single amino acid, a whole region is in the centre of interest. Examples are functional domains, like the ligand-binding domain or the channel domain, and segments critical for protein signalling, sorting or targeting.

A common way to designate a receptor chimera is to use the names of the receptor types and the name of the joining amino acid (*e.g.*, alpha7-V201-5HT$_3$ refers to a chimeric receptor with the N-terminal part of the α7 nAChR joined with the C-terminal part of the 5HT$_3$ receptor at the amino acid valine 201.

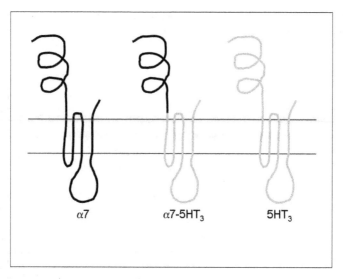

Fig. 2. Schematic representation of a single subunit each from the α7 receptor, the a7-5HT₃ chimera, and the 5HT₃ receptor. The black line denotes the α7 protein; the grey line denotes the 5HT₃ protein.

1.3 In vitro systems to test functional properties of mutated receptors

Electrophysiology is a critical tool for receptor research. This section gives a basic introduction to the reader not familiar with this technique. Two methods are commonly used: (1) the two-electrode voltage clamp method using *Xenopus* oocytes and (2) the whole-cell voltage clamp method amenable for cell lines.

1. Oocytes from the frog *Xenopus laevis* produce almost all ion channel receptor types in high amounts upon injection of their mRNAs. Since only transient protein expression is possible in this system, oocytes can be used only for a short period of a few days. As the name suggests, two sharp microelectrodes filled with a high molarity potassium chloride solution are pricked into 0.6 to 1 mm oocytes to initiate the recording of channel activity. Fortunately an oocyte can be used for several hours. This is enough time to allow the study of a series of agonists and antagonists.
2. The whole-cell patch-clamp technique is the most common electrophysiological method used for many mammalian cell lines such as HEK-293, CHO, GH4C1 or PC-12 cells. These cells can be easily transfected with nAChR expression vectors. It is possible to generate stable cell lines that have integrated the nAChR sequence into the genomic DNA. The whole-cell voltage clamp method employs thin pipettes, which enable good electrical access to the interior of a cell combined with full external electrical insulation. This technique allows measuring small electric currents generated by ion flow through a single receptor molecule or through a couple of receptors. The fast gating and desensitizing channels require fast and direct drug application methods, such as U-tubes and Y-tubes. These systems can only be used with the small mammalian cells (10

– 20 μm). Heterologously expressed receptors in mammalian cell lines might differ in their biophysical and pharmacological characteristics as compared to those analyzed in *Xenopus laevis* oocytes. Therefore, results from the two systems are not always in agreement.

2. Lessons learned from directed mutagenesis

2.1 Identification of intramolecular signals for nicotinic receptor targeting

The physiological role of nAChRs depends on their localisation in specific regions of the cell. For example, presynaptic receptors regulate the transmitter release from synaptic vesicles into the synaptic cleft. Postsynaptic receptors modulate the postsynaptic potential that stimulates action potential formation in neurons or at the neuromuscular junction (Wonnacott et al., 1997; Albuquerque et al., 2009). The observation of differential expression of nAChRs at pre- and postsynaptic sites triggered the search for specific cellular localization signals. This section illustrates mutagenesis approaches to investigate receptor targeting.

Chicken ciliary neurons proofed to be an excellent tool to study receptor targeting (Williams et al., 1998; Temburni et al., 2000). Ciliary ganglion neurons express two nAChR subtypes: (1) α7 nAChRs, for which localisation is restricted to the perisynaptic dendritic membrane and (2) heteromeric nAChRs consisting of α3, α5 and β4 subunits, which are expressed primarily in postsynaptic membranes (Jacob et al., 1986; Conroy and Berg, 1995). The α3 and α7 nAChR subunits are highly homologous to each other, with one exception: the long cytoplasmatic loop shows great diversity in sequence and length (Lindstrom et al. 1996). It also has been shown that this loop is required for the cellular sorting and trafficking machinery. Therefore, it might be the candidate domain for subcellular targeting. Chimeric α7 nAChR subunits were constructed, in which the cytoplasmatic loop was replaced by the homologous region of the α3 nAChR subunit. Furthermore, a myc-epitope tag was added at the C-terminus to allow detection of the receptor without affecting function. Then the chimaeric α7 nAChR subunit with the α3 nAChR cytoplasmatic loop was ectopically expressed in chicken ciliary neurons. Indeed this chimeric receptor was targeted to the postsynaptic membrane, as shown by antibody staining of the myc-epitope tag. This result demonstrates that the cytoplasmatic loop of the α3 nAChR governs the subcellular targeting of the receptor. Other endogenous nAChR subunits do not play this role because the nAChR subunit does not co-assemble with the α3 nAChR subunit (Conroy and Berg, 1995). In addition, α7 nAChR chimeric subunits containing the cytoplasmatic loop of the α5 or β4 nAChR subunit were designed and expressed. These chimeras were targeted to the perisynaptic site (Temburni et al., 2000). This result means that not only the α7 nAChR subunit contains signals for perisynaptic localisation, but that the α5 and β4 nAChR subunits also do.

Interestingly, in the case of the α3α5β4 heteromer, in which perisynaptic and postsynaptic signals are both present, it is the cytoplasmatic loop of the α3 nAChR subunit that determines the targeting to the postsynaptic site of ciliary ganglion neurons. Taken together the results show that the cytoplasmatic loop contains the cellular localisation signal.

The next step was to determine the exact signal peptide within the cytoplasmatic loop. Before initiating the studies, it was tried to transfer the described approach to hippocampal neurons (Xu et al., 2006). Again the cytoplasmatic loops between the α4 and α7 nAChR subunits were swapped and ectopically expressed in various combinations in hippocampal neurons. Unfortunately, no surface expression could be detected in cells expressing these chimeric nAChR subunits, either alone or in various combinations. It was possible that the design of the chimera was too aggressive. Possibly a critical peptide sequence needed to enable proper receptor assembly and expression in hippocampal cells was accidentally removed.

Therefore, another mutagenesis strategy was chosen. Swapping internal protein domains may affect receptor assembly. Instead model proteins were used. They were left intact, but they were tagged with putative signal peptides from the nAChR. Two non-neuronal transmembrane proteins were chosen. CD4 and the Interleukin 2 receptor when heterologously expressed in neurons are evenly distributed (Gu et al., 2003). It was tested whether the intracellular loops of the α4 and α7 nAChR subunits are able to target these proteins to specific sites in neurons. When the cytoplasmatic loop of the α7 nAChR was fused to the intracellular-oriented C-terminus of CD4, the chimeric protein was only detected in the dendrites. In contrast, the homologous α4 nAChR cytoplasmatic loop leads to axonal expression.

In order to narrow down the precise localisation signals in the cytoplasmatic loop of the nAChR subunits, the following strategy was chosen. Various overlapping fragments covering the loop from its N-terminal region to its C-terminal region were fused to the C-terminus of the Interleukin-2 receptor. The chimeric receptors were expressed in hippocampal neurons. A specific 25-amino acid-fragment (residue positions 30-54) of the α4 nAChR cytoplasmatic loop targeted the chimera to axons. A 48-residue fragment (positions 33-80) of the α7 nAChR cytoplasmatic loop targeted the chimaera to dendrites (Xu J. et al. 2006).

In a last step, site-directed mutagenesis of specific amino acid residues identified in the targeting sequences a leucine motif (DEXXXLLI) in the α4 nAChR cytoplasmatic loop and a tyrosine motif (YXXx) in the α7 nAChR loop.

In conclusion, an iterative approach of chimera design has pin-pointed the precise targeting sequence of a receptor. It is important to note that chimera design may destroy receptor expression or function. Therefore, it is advisable to generate a set of chimeras and to recognize that the change of an expression system may require an adaption of chimera design.

2.2 Confirming computer-based predictions for posttranslational modifications

Posttranslational modifications are important mechanisms for regulating protein expression and protein activity in eukaryotes. Three posttranslational modifications are known for the nicotinic acetylcholine receptor family: glycosylation, phosphorylation and palmitoylation (Albuquerque et al., 2009). Modern computational algorithms effectively help to identify sequence motifs for putative posttranslational modifications. However, experimental approaches are needed to confirm that the site is actually used for posttranslational modifications and to understand its physiological role. By illustrating the role of phosphorylation, this section exemplifies how to combine computational tools with mutagenesis strategies.

Phosphorylation of α7 nAChR negatively regulates its activity. For example, tyrosine kinase inhibition by genistein, a kinase inhibitor, decreases α7 nAChR phosphorylation and, as a consequence, strongly increases acetylcholine-evoked currents (Charpantier et al., 2005). Therefore, it is interesting to know where and which phosporylation sites are present in the protein sequence. Computer analysis predicts two putative phosphorylation sites (tyrosines at residues 386 and 442) in the long cytoplasmatic loop between TM3 and TM4 of the human α7 nAChR. A site-directed mutagenesis that replaced the tyrosines with alanines (Y386A and Y442A) was carried out, and a receptor double mutant with both mutated phosphorylation sites was tested in *Xenopus* oocytes. Indeed the activity of the receptor double mutant was increased to an extent comparable to inhibition of the wild-type receptor by genistein (a kinase inhibitor). This result confirmed that at least one of the two sites is a physiological phosphorylation site. As the receptor double mutant was insensitive to genistein, it can also be assumed that there are no additional physiologically relevant phosphorylation sites in the protein sequence, which might have been overlooked by the computer algorithm. We note that the value of the receptor mutant not only lies in confirming the phosphorylation site, but also in establishing a physiological role for this posttranslational modification. It is primarily required for the regulation of receptor activity, rather than for receptor expression.

2.3 Receptor chimeras demonstrate the modular domain structure of nAChRs

As mentioned in the introduction, nAChRs are integral signal transducers in which the signalling domain and the response element are within one protein. A key question for understanding the molecular design of a receptor family is whether this is achieved through a modular architecture with functionally independent and separable elements.

In 1993 it was impressively demonstrated that domains are interchangeable between different ligand-gated ion channels. This was exemplified with chimeras made from the α7 nAChR and the 5HT$_3$ receptors. The 5HT$_3$ receptor is modulated by the ligand serotonin (5HT), permeable for Na$^+$ and K$^+$ ions, but it is blocked by Ca^{2+} ions. The authors constructed recombinant chimeric receptors with the N-terminal part of the α7 type nAChR, containing the ligand-binding site, and the C-terminal part of the 5HT$_3$ receptor, containing the ion channel domain. They constructed five different chimeras with different junction points for the two receptor parts, using conserved residues W173, Y194, V201, L208 and P217 (the numbers refer to the residue of the α7 receptor). Four of the five chimeric subunit constructs produced properly assembled membrane receptors with an intact extracellular ligand-binding domain, as confirmed by radioactive α-bungarotoxin-binding assays. (α–bungarotoxin is a competitive inhibitor of acetylcholine and binds with high affinity to the acetylcholine-binding site.) Two of the five chimaeras were functional, as shown by the two-electrode voltage clamp technique in *Xenopus* oocytes. One chimera (V201) was able to form large acetylcholine-evoked currents; another (Y194) was able to form small currents.

The V201 chimera was obviously the most interesting chimera and, therefore, was investigated thoroughly by electrophysiological methods and compared to the respective wild-type receptors.

First the ligand-binding properties of the chimaera were pharmacologically studied with a set of modulators and ligands. The agonists were acetylcholine and nicotine, and the antagonists were α-bungarotoxin and curare, both of which have little or no effect on the $5HT_3$ receptor. In contrast 5HT, the natural ligand of the $5HT_3$ receptor, has no effect on the α7 nAChR receptor. In the *Xenopus* oocyte system, the chimera responded to these ligands in a manner similar to the response of the α7 nAChR wild-type receptor, *i.e.*, acetylcholine increased the current; curare inhibited the currents; and 5HT had no effect.

Second the ion channel properties of the chimera were investigated. α7 type nAChRs and $5HT_3$ receptors differ in their sensitivity to external calcium. While α7 type nAChRs currents increase with higher calcium concentrations, $5HT_3$ receptor currents decrease. The α7 nAChR channel domain is highly conductive for calcium ions, whereas the $5HT_3$ receptor channel domain is blocked by calcium. In addition, external calcium ions have a potentiating effect of the acetylcholine action on α7 nAChRs. In *Xenopus* oocyte studies, the α7 nACh-V201-$5HT_3$ receptor chimera behaved in a manner similar to the response of the $5HT_3$ receptor towards the external calcium concentration. This means that the ion channel properties of the $5HT_3$ receptor are independent of its ligand-binding properties. It appears that, by swapping the ligand-binding domains, a $5HT_3$ receptor can be engineered to respond to acetylcholine and other nAChR modulators like a natural nAChR.

Interestingly, the onset and desensitization kinetics of the current of the chimera is in between the rapid kinetics of wild-type α7 nAChR and the slow kinetics of the $5HT_3$ receptor. This suggests that not all properties of a receptor can be exchanged by simply swapping the ligand-bind domains. The specific kinetic properties of a receptor apparently require interplay between different domains.

In conclusion, computer-based predictions and traditional biochemistry can certainly suggest the domain structure of a protein. However, in order to distinguish whether domains are functionally independent or just building blocks of a larger functional unit, chimeras are invaluable tools.

2.4 Concatemeric nACh receptors revealed the subunit order of nAChRs

The principal architecture of heteromeric nAChRs allows for, at least theoretically, numerous combinatorial arrangements of the various α, β, γ, δ and ε subunit types around the central pore. These arrangements could result in many different receptors with different properties. Which of the possible arrangements are realized in nature, and what distinct properties might they have? For example, it is conceivable that different arrangements of the subunits have different acetylcholine-binding properties because the binding site is located at the interface of two subunits.

In neuronal tissue, many different nAChR combinations are usually present. It can be difficult to study individual nAChR types. In addition the actual arrangement of the subtypes cannot easily be assessed. Directed mutagenesis can be used to force subtypes to form specific nAChR arrangements. Thus, it can be a powerful tool for investigating the roles of different nAChR arrangements. This section focuses on the α4β2 type nAChR, which is the most abundant form of heteropentameric nACh receptors in the mammalian brain.

A pentameric α4β2 nAChR could consist of three α4 and two β2 subunits, or vice versa. Earlier research using *Xenopus laevis* oocytes has shown that the use of equal amounts of α4 and β2 subunit mRNAs generates an $(\alpha4)_2(\beta2)_3$ nAChR (*i. e.*, the α4 and β2 subunits are in a 2:3 stoichiometry, respectively) (Anand et al., 1991; Cooper et al., 1991). In contrast, under conditions in which the α4 subunit mRNA was in excess, an $(\alpha4)_3(\beta2)_2$ nAChR was obtained (3:2 stoichiometry) (Zwart and Vijverberg, 1998). The two different α4β2 nAChRs differ in their pharmacological properties. The $(\alpha4)_2(\beta2)_3$ nAChR is highly sensitive to acetylcholine, whereas the $(\alpha4)_3(\beta2)_2$ nAChR is less sensitive. Since both receptor forms are present in the brain (Marks et al., 1999; Gotti et al., 2008), it has been suggested that the ratio of the two nAChRs is part of a regulatory mechanism for neuronal cell response to nicotine (Tritto et al., 2002; Kim et al., 2003). In addition to their differential sensitivity to acetylcholine and nicotine, the two receptor forms differ in other properties, such as desensitization kinetics and Ca^{2+} permeability (Nelson et al., 2003; Zwart et al., 2008; Moroni et al., 2006; Tapia et al., 2007; Moroni et al., 2008).

For each stoichiometric nAChR, there exist two possible orders of the subunits. The $(\alpha4)_2(\beta2)_3$ nAChR can be α4–α4–β2–β2–β2 (1.1) or α4–β2–α4–β2–β2 (1.2). The $(\alpha4)_3(\beta2)_2$ nAChR can be α4–α4–α4–β2–β2 (2.1) or α4–α4–β2–α4–β2 (2.2). Which of the possible arrangements are realized and what functional significance is conferred by specific positions of each subunit within the receptor complex? Insight into this issue has been gained by using a specific type of directed mutagenesis, *i.e.*, concatemers. A concatemer is a long synthetic gene that contains smaller genes (*e.g.*, the subunit genes) linked in series. In this case, the resulting protein consists of a defined sequence of subunits in which the carboxyl-terminus of the preceding subunit is covalently linked with the amino-terminus of the following subunit (Zhou et al., 2003; Nelson et al, 2003). This technique allows one to enforce a predefined subunit order for the receptor. Studies employing this technique with tandem and triple concatemers of the α4 and β2 subunits showed that only two arrangements were functional. For the $(\alpha4)_2(\beta2)_3$ nAChR, it was the α4–β2–α4–β2–β2 (1.2) arrangement; and for the $(\alpha4)_3(\beta2)_2$ nAChR, it was the α4–α4–β2–α4–β2 (2.2) arrangement (Carbone et al., 2009). Thus, in both receptor forms a triplet of the same subunit was avoided. In summary, the mutagenesis studies described improved our understanding of how the subunit types are arranged (Zhou et al., 2003; Nelson et al, 2003).

Recently, the use of a pentameric nAChR concatemer revealed that there exists also at the α4/α4 interface a functional acetylcholine-binding site (Mazzaferro et al., 2011). Originally, the acetylcholine-binding sites were thought to be located at the α4/β2 subunit interfaces. As these interfaces are present in both receptor isoforms, it is unlikely that they account for differences in acetylcholine sensitivities. In the new study the authors clearly identified the α4/α4 interface in α4–α4–β2–α4–β2 receptors as an additional acetylcholine-binding site. They used a combined approach of a pentameric nAChR concatemer, chimeric α/β subunits with mutagenesis of loop C and structural modelling to determine that this α4/α4 interface accounts for isoform-specific characteristics, *i.e.*, for the low acetylcholine sensitivity. In conclusion, directed mutagenesis permitted a defining of the order of the nAChR subunits and, thus, allowed a determination of the way agonist-binding sites are formed.

2.5 Cobratoxin

The α-neurotoxins from snake venoms are potent antagonists of nicotinic acetylcholine receptors. In mouse some of them are over ten times more toxic than nicotine (*e.g.*, the LD_{50}

of α-cobratoxin from *Naja naja* in mouse is 0.4 mg/kg versus 7.1 mg/kg for nicotine). Despite the high toxicity of the snake toxins, some animals are resistant to α-neurotoxins. This is the case for animals that feed on cobras, such as the mongoose, and, of course, the snake itself. A substantial effort using directed mutagenesis has been made to characterize the interaction between α-neurotoxins and the nicotinic acetylcholine receptors to identify the mechanism of this resistance.

2.5.1 Understanding the cobratoxin – α7 nicotinic acetylcholine receptor interaction

The binding site of the α7 nAChR is composed of six loops, A to F. In order to identify which of these loops interact(s) with α-cobratoxin (α-Cbtx), extensive site-directed mutagenesis was carried out to generate 40 receptor mutants (Fruchart-Gaillard et al., 2002). The possible role of a given amino acid in a α7 nAChR loop thought to interact with α-Cbtx was determined by comparing a mutant receptor with a changed amino acid to the wild-type receptor using a competition binding assay with radioactive iodide-labelled α-bungarotoxin. Only mutations in loops C, D and F reduced the affinity to α-Cbtx. Hence, these loops may be critical for α7 nAChR - α-Cbtx interaction. Mutations of loop C at residues F186 and Y187 showed the greatest effect. They reduced the affinity by 100- to 200-fold. It is important to notice that not every mutation at these two positions reduced affinity. For example, F186R reduced affinity by a factor of 100; in contrast, F186A reduced affinity by a factor of only 4; and F186T, not at all. Hence, when determining a possible role of an amino acid by site-directed mutagenesis, amino acid properties may matter (*e.g.*, size or charge). It is often important to test several amino acid exchanges for a full understanding.

Similarly, a site-directed mutagenesis approach was taken to generate 36 toxin mutants. The objective was to identify the interaction sites of α-Cbtx with the receptor (Antil-Delbeke et al., 2000). This study found loop II and the C-terminal tail of α-Cbtx to interact with α7 nAChR.

To determine which amino acid(s) of the receptor interacts with which amino acid(s) of the toxin, α7 nAChR receptor mutants were tested with α-Cbtx mutants in the competition assay. The studies revealed that the amino acid R33 in loop II of α-Cbtx interacts with a number of amino acids in loop C of the α7 nAChR, such as Y187, W148, P193, and Y194 (Fruchart-Gaillard et al., 2002). Another example is the amino acid K35 of α-Cbtx that interacts with the amino acids F186 and D163 of the α7 nAChR.

This information was then used in a computational 3D model to orient α-Cbtx in the binding pore of α7 nAChR and to help understand the mechanism of the antagonistic action of α-Cbtx. How can the large α-Cbtx molecule exert its antagonistic action on a binding site that is configured to fit small ligands, such as acetylcholine or nicotine? The docking study revealed that only the tip of loop II of the toxin plugs into the cavity between two receptor subunits. About 75% of the remaining surface of the toxin stays outside the toxin-receptor complex. In this way, α-Cbtx behaves like a small ligand and effectively antagonizes α7 nAChR.

2.5.2 Establishing resistance against snake toxin

A snake is usually resistant to its own venom. Hence, it is interesting to find out whether it is resistant from a specific difference in the target molecule of its toxin. In this case, the target molecule is the nicotinic acetylcholine receptor. There are considerable differences in the protein sequences of the nAChR ligand-binding domains of snakes (*Naja spes*) and mammals. The differences may suggest the different sensitivities. It is possible to make the snake α1 nAChR sensitive to α-Bungarotoxin, a venom of the elapid family. It has been shown by site-directed mutagenesis that introducing the mutation N189F, which substitutes the asparagine in the snake protein sequence with the phenylalanine in the mouse sequence, abolishes resistance (Takacs et al., 2001). This suggests not only that the snake α1 nAChR contains a ligand-binding domain for snake toxins, but also that a single amino acid can cause sensitivity or resistance. In order to test whether the asparagine indeed confers the resistance, a F189N mutation was introduced into the nAChR of the mouse. Two-electrode voltage clamp analysis in *Xenopus* oocytes showed that this mouse receptor mutant was resistant to α-Bungarotoxin. This means that a single amino acid substitution can determine sensitivity or resistance to a snake venom.

Interestingly, N189 is an N-glycosylation site, and it has been postulated that the bulky glycosyl residue may prevent the toxin from entering the binding site (Barchan et al., 1992). This hypothesis might explain the resistance of the mongoose to snake venoms and the sensitivity of mammals. In fact, the mongoose nAChR has an N-glycosylation site at N187, whereas other mammals, such as mouse, cat and humans, lack an N-glycosylation site in the ligand-binding domain.

Another resistance mechanism may be deduced from the receptor-toxin interaction, as described in the previous section. The F189 residue in the α1 nAChR is analogous to F186 in the α7 nAChR. F186 was identified to be critical for the receptor-binding interaction. Therefore, it is conceivable that two mechanisms in the snake receptor confer resistance to α-neurotoxins: (1) disruption of a critical protein-protein interaction and (2) steric hindrance via glycosylation.

2.6 Mapping of the binding sites of allosteric potentiating ligands

Modulators of ligand-gated ion channels (LGICs) have become therapeutically important because, in contrast to traditional agonists or antagonists, these substances change receptor activity only in the presence of the natural ligand. This allows a more physiological control of a LGIC. This concept has been broadly applied for the treatment of epilepsy. For example, benzodiazepines, such as diazepam ("Valium"), present a major class of allosteric modulators of the $GABA_A$-receptors.

Cholinergic neurotransmission is a prominent therapeutic target for the treatment of diseases like Alzheimer's disease. The drug galantamine exerts its therapeutic action by allosteric modulation of the nAChRs (Bertrand and Gopalakrishnan, 2007; Maelicke et al., 2001). Understanding the mechanism of allosteric modulation is therefore important for developing novel drugs for treating Alzheimer´s disease (Faghih et al., 2007).

Today a range of allosteric modulators is known. Synonymously used terms are "allosteric potentiating ligands" (APL) and "positive allosteric modulators" (PAM). They fall into different classes. Galantamine is a representative of the type I class of PAMs (PAM I), which enhance nAChR activity by increasing the current without affecting receptor desensitization. In contrast, members of the type II class of PAMs (PAM II) increase the current of nAChRs, but also reduce their desensitization (Hurst et al., 2005). Hence, a number of studies were directed to the identification of possible different binding sites, which would help to explain the mechanistic differences of the two classes.

2.6.1 PNU-120596

The group around Neil Millar ran a combined approach to locate the binding site of the type II modulator PNU-120596 a well-studied developmental compound. In a first round of experiments, they compared the action of PNU-120596 on the α7 nAChR and the $5HT_3$ receptor with a set of α7 nACh/$5HT_3$ receptor chimeras (Young et al., 2008). In these chimeras, principally the extracellular part of the α7 nAChR was combined with at least the first three transmembrane domains from the $5HT_3$ receptor. PNU-120596 could, of course, potentiate α7 nAChR. However it could potentiate neither the original $5HT_3$ receptor nor any of the receptor chimeras. This indicated that the PNU-120596 binding site is located in the transmembrane part of the nACh receptor.

In a second round of mutant receptor designs, the amino acid sequence of the transmembrane part of the α7 nAChR was compared with the one from the $5HT_3$ receptor. A number of differences were identified. Amino acids that are not conserved in the α7-nAChR (which is potentiated by PNU-120596) and the $5HT_3$ receptor (which is not potentiated by PNU-120596) formed the basis for a set of α7 nAChR mutants. Amino acids in the nAChR were mutated to the corresponding amino acids of the $5HT_3$ receptor. As expected, some of the mutants were simply not functional, whereas others showed responses to PNU-120596 that were similar to that of wild-type nAChR. Five mutants having otherwise normal function were less responsive than wild type to the potentiating activity of PNU-120596. Two of them (A225D and M253L) were nearly resistant to PNU-120596.

In order to understand why these five amino acids of wild-type nAChR conferred sensitivity to PNU-120596 modulation, the authors investigated several computer models of the nAChR. It turned out that the five amino acids were part of an intra-subunit cavity. Docking simulations revealed that the most favourable docking position of PNU-120596 was at a location very near to the locations of the five amino acids.

Complementary studies using a set of different α7/5HT3 receptor chimeras showed that the α7 nAChR ion channel domain is essential for the action of PNU-120596 (Bertrand et al., 2008). This has led to the model that PNU-120596 exerts its function by stabilizing the cavity in an agonist-like fashion (Barron et al., 2009).

In summary, a combined approach of site-directed mutagenesis experiments, molecular modelling and docking studies was needed to identify the binding site of PNU-120596.

2.6.2 Galantamine

The following studies revealed a totally different location for the binding site of galantamine. They offer an explanation of why the mechanisms of action of the two allosteric modulating ligands are substantially different.

The identification of the galantamine binding site on the extracellular domain close to the acetylcholine binding site has taken more than a decade. Several different approaches were required to finally locate it precisely (Schröder et al., 1994; Ludwig et al., 2010).

A first step was the combination of results from earlier studies about an antibody called FK1 (Schröder et al., 1994; Brejc et al., 2001; Luttmann et al., 2009). An important feature of antibody FK1 is its ability to block the potentiating effect of galantamine on the nAChR without affecting the response to acetylcholine or agonists. The strategy described in this paper helped to identify two stretches (27 and 28 amino acids in length) that contain part of the binding site of galantamine. However, the question remained as to which amino acids participate in the interaction (Schröder et al., 1994). A protein from a freshwater snail assisted in narrowing the range of possible amino acids (Brejc et al., 2001; Smit et al., 2001). The freshwater snail *Lymnaea stagnalis* produces a protein that is homologous to the ligand-binding domain of nAChRs. It is called Acetyl Choline Binding Protein, and it assembles as a homopentamer amenable to X-ray crystallography. A 2.7 Å resolution structure was used as a template to model the ligand-binding domains of α7 and α4β2 nAChRs (Luttmann et al., 2009). In these models only a small proportion of the amino acid stretches of the receptor lie in a position at the outer surface that would be accessible to the FK1 antibody (Figure 3). As both amino acid stretches contribute to the FK1 epitope, it seems highly probable that the epitope is located at the junction of the two stretches.

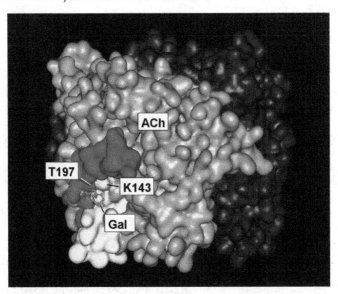

Fig. 3. Surface model of the α7 nAChR ligand-binding domain. Only the ligand-binding domain (LBD) is shown. The cell membrane and channel domain would be beneath the LBD. Two of the five subunits are shown in light grey, with the other ones depicted in dark grey. One molecule of acetylcholine (ACh) is bound. It is almost buried inside the binding site. The amino acids in yellow and blue belong to amino acid stretches identified in Schröder et al., 1994 as contributing to the epitope for the galantamine blocking antibody FK1. Mutation of the amino acids T197 and K143 showed no (T197) or reduced (K143) galantamine effect when stimulated with acetylcholine. This is in line with the assumption that the galantamine binding site is at the junction of these two amino acid stretches. The amino acids T197 and K143 are possible binding sites for galantamine (Gal), as predicted by docking studies.

These insights led to a hypothetical binding site, which was proven with a set of eight different α7 nAChR mutants (Ludwig et al., 2010). All of the mutants were functional, albeit two had much lower affinities for acetylcholine and agonists than wild type. Four of the mutants, all of which showed a normal response to agonists, had no or a reduced response to galantamine. These four mutants were altered in amino acids at the borders of the two stretches mentioned above and were adjacent in the α7 nAChR model (Ludwig et al., 2010). Docking studies pinpointed two amino acids, threonine 197 and lysine 143, as the galantamine binding site (Luttmann et al., 2009). The two amino acids that were replaced in the other two mutants were shown to be oriented in a way in which they would be unlikely to interact with galantamine directly.

A mechanistic model to explain how the binding of acetylcholine opens the ion channel proposes that the C-loop (Fig. 1) acts as a lever that moves upon binding of acetylcholine. Since the galantamine binding site is located at the lower part of the C-loop, it is assumed that binding of galantamine enhances the action of this lever.

2.6.3 Ivermectin

Ivermectin is a member of the PAM I class of modulators. It has a potentiating action on the nAChR activation without affecting the desensitization. Collins and Millar performed an approach similar to the one described for PNU-120596 to identify the amino acids that play a critical role in the interaction with ivermectin (Collins and Millar, 2010).

The authors conducted experiments with α7 nAChRs, 5HT$_3$ receptors and the already described α7-5HT$_3$ chimeric receptors in *Xenopus* oocytes. The experiments led to the conclusion that the transmembrane domain plays a critical role in the allosteric modulation by ivermectin. The results showed that, while ivermectin potentiated the effect of acetylcholine on the α7 nAChR, it had no effect on the 5HT$_3$ receptor. In contrast, the α7-5HT$_3$ chimeric receptor, which contains the ligand-binding domain of the α7 nAChR and the ion channel domain of the 5HT$_3$ receptor, was surprisingly inhibited by ivermectin. The reason for this unexpected response of the chimeric receptor to ivermectin is not known, but it might be that the 5HT$_3$ receptor extracellular domain blocks the access of ivermectin to a transmembrane domain binding site that is accessible in the chimeric receptor.

In subsequent experiments, the authors changed selected amino acids in the α7 nAChR. The mutations A225D, Q272V, T456Y, and C459Y almost completely prevented allosteric modulation by ivermectin. It is interesting to note that some of these mutants react similarly to PNU-120596 and to ivermectin. The A225D and C459Y mutants showed a reduced response to both compounds. In contrast, mutations at Q272V and T456V reduced the allosteric potentiation of ivermectin, but not of PNU-120596. This suggests that the amino acids responsible for the allosteric potentiating action of ivermectin partially overlap the ones responsible for the action of PNU-120596.

2.6.4 Conclusion

Section 2.6 describes the use of directed mutagenesis for the identification and investigation of the binding sites of the three nAChR modulators PNU-120596, galantamine and ivermectin. Galantamine and ivermectin are classified as PAM I modulators, while PNU-120596 is a PAM II class allosteric modulator, based on electrophysiological properties. However, the binding site of galantamine is on the extracellular ligand-binding domain, while ivermectin and PNU-120596 bind to the channel domain of the receptor. Apparently similar electrophysiological properties do not reflect similar binding sites.

3. Concluding remarks

This chapter has highlighted milestones of the directed mutagenesis research performed on nAChRs. It includes research about receptor targeting, the modular domain architecture, the order of the heteromeric subunit assembly, the mechanism of toxin resistance and the receptor interaction with modulatory ligands. Future research will continue to investigate the nicotinic receptor family and its role in diseases, such as Alzheimer's or Parkinson's. The interaction between α7 nAChR and β-amyloid is just one of the many urgent questions that need to be resolved (Tong et al., 2011). Directed mutagenesis will remain one of the most powerful tools on the journey towards the full understanding of this molecular machine.

4. References

Anand R., Conroy, W.G., Schoepfer, R., Whiting, P., & Lindstrom, J. (1991). Neuronal nicotinic acetylcholine receptors expressed *Xenopus* oocytes have a pentameric quaternary structure. *Journal of Biological Chemistry*, Vol.266, No.17 (June 1991), pp. 11192-11198, ISSN 0021-9258.

Antil-Delbeke, S., Gaillard, C., Tamiya, T., Corringer, P., Corringer, P., Servent, D., & Ménez, A. (2000). Molecular determinants by which a long chain toxin from snake venom interacts with the neuronal alpha7-nicotinic acetylcholine receptor. *Journal of Biological Chemistry* Vol.275, No.38 (September 22), pp.29594–29601 ISSN 1083-351X

Albuquerque, E., Pereira, E., Alkondon, M., & Rogers, S. (2009). Mammalian nicotinic acetylcholine receptors: from structure to function. *Physiological Reviews*, Vol.89, No.1, (January 2009), pp. 73-120. ISSN 1522-1210

Barchan, D., Kachalsky, S., Neumann, D., Vogel, Z., Ovadia, M., Kochava, E., & Fuchs, S. (1992). How the mongoose can fight the snake: The binding site of the mongoose acetylcholine receptor. *PNAS* Vol. 89, No.16 (August 15), pp.7717-7721, ISSN 0027-8424

Barron, S.C., McLaughlin, J.T., See, J.A., Richards, V.L., & Rosenberg, R.L., (2009). An allosteric modulator of alpha7 nicotinic receptors, N-(5-Chloro-2,4-dimethoxyphenyl)-N'-(5-methyl-3-isoxazolyl)-urea (PNU-120596), causes conformational changes in the extracellular ligand binding domain similar to those caused by acetylcholine. *Molecular Pharmacology*, Vol.76, No.2 (August 2009), pp. 253-263, ISSN 0026-895X

Bertrand, D., Bertrand, S., Cassar, S., Gubbins, E., Li, J., & Gopalakrishnan, M. (2008). Positive allosteric modulation of the alpha7 nicotinic acetylcholine receptor: ligand interactions with distinct binding sites and evidence for a prominent role of the M2-M3 segment. *Molecular Pharmacology*, Vol.74, Nr.5, (November 2008), pp.1407-1416, ISSN 0026-895X

Brejc, K., van Dijk, W., Klaassen, R., Schuurmans, M., van Der Oost, J., Smit, A. and Sixma, T. (2001). Crystal structure of an ACh-binding protein reveals the ligand-binding domain of nicotinic receptors. *Nature*, Vol.411, No.6835, (May 2001), pp.269-276, ISSN 0028-0836

Carbone, A.L., Moroni, L.M., Groot-Kormelink, P.J. & Bermudez, I. (2009). Pentameric concatenated $(\alpha 4)_2(\beta 2)_3$ and $(\alpha 4)_3(\beta 2)_2$ nicotinic acetylcholine receptors: subunit arrangement determines functional expression. *British Journal of Pharmacology*, Vol. 156, No.6 (March 2009), pp. 970–981, ISSN 0007-1188.

Charpantier, E., Wiesner, A., Huh, K., Ogier, R., Hoda J., Allaman, G., Raggenbass, M., Feuerbach, D., Bertrand, D., & Fuhrer, C. (2005). $\alpha 7$ Neuronal nicotinic acetylcholine receptors are negatively regulated by tyrosine phosphorylation and src-family kinases. *The Journal of Neuroscience*, Vol.25, No.43, (October 26), pp.9836 – 9849, ISSN: 1529-2401

Collins, T., and Millar, N.S. (2010). Nicotinic acetylcholine receptor transmembrane mutations convert ivermectin from a positive to a negative allosteric modulator. *Molecular Pharmacology*, Vol.78, No.2, (August 2010), pp. 198-204, ISSN 0026-895X

Conroy, W.G., & Berg D.K. (1995). Neurons can maintain multiple classes of nicotinic acetylcholine receptors distinguished by different subunit compositions. *The Journal of Biological Chemistry*, Vol.270, No. 9, (March 1995), pp. 4424- 4431, ISSN 0021-9258

Cooper, E., Couturier, S. & Ballivet, M. (1991). Pentameric structure and subunit stoichiometry of a neuronal nicotinic acetylcholine receptor. *Nature*, Vol.350, No.6315 (March 1991), pp. 235-238, ISSN 0028-0836.

Drisdel, R., Manzana, M. & Green, W (2004). The role of palmitoylation in functional expression of nicotinic $\alpha 7$ receptors. *The Journal of Neuroscience*, Vol.24 No.46, (November 17), pp. 10502–10510, ISSN: 1529-2401

Eiselé, J., Bertrand, S., Galzi, J., Devillers-Thiéry, A., Changeux J., & Bertrand D. (1993) Chimaeric nicotinic-serotonergic receptor combines distinct ligand binding and channel specificities. *Nature*, Vol.366, No.6454, (December 1993), pp.479-483, ISSN: 0028-0836

Fruchart-Gaillard, C., Gilquin, B., Antil-Delbeke, S., Le Novère, N., Tamiya, T., Corringer, P., Changeux, J., Me'nez, A., & Serven, D. (2002). Experimentally based model of a complex between a snake toxin and the $\alpha 7$ nicotinic receptor. *PNAS*, Vol.99, No.5, (March 5), pp.3216–3221, ISSN 0027-8424

Faghih, R., Gfesser, G., & Gopalakrishnan, M. (2007). Advances in the discovery of novel positive allosteric modulators of the alpha7 nicotinic acetylcholine receptor. *Recent Patents on CNS Drug Discovery*, Vol.2, No.2 (June 2007), pp. 99-106, ISSN 1574-8898

Gahring, L., & Rogers, S. (2006). Neuronal nicotinic acetylcholine receptor expression and function on nonneuronal cells. *AAPS Journal*, Vol.7, No.4, (January 2006), pp. E885-894, ISSN ISSN 1550-7416

Geerts, H., Guillaumat, P., Grantham, C., Bode, W., Anciaux, K., & Sachak, S. (2005), Brain levels and acetylcholinesterase inhibition with galantamine and donepezil in rats, mice, and rabbits. *Brain Research*, Vol.1033, No.2, (February 2005), pp. 186-193, ISSN 0006-8993

Gotti, C., Zoli, M., & Clementi, F. (2006). Brain nicotinic acetylcholine receptors: native subtypes and their relevance. *Trends In Pharmacological Science*, Vol.27, No.9 (September 2006), pp. 482–491, ISSN 0165-6147.

Gotti C., Clementi F., Fornari A., Gaimarri A., Guiducci S., Manfredi I., Moretti M., Pedrazzi P., Pucci L., & Zoli M. (2009). Structural and functional diversity of native brain neuronal nicotinic receptors. *Biochemical Pharmacology*, Vol.78, No.7, (October 2009), pp. 703-711. ISSN 0006-2952

Hucho, F., Tsetlin, V. & Machold, J. (1996). The emerging three-dimensional structure of a receptor. The nicotinic acetylcholine receptor. *European Journal of Biochemistry*, Vol.239, No.3, (August 1996), pp.539-557, ISSN 1432-1033

Hurst, R., Hajós, M., Raggenbass, M., Wall, T., Higdon, N., Lawson, J., Rutherford-Root, K., Berkenpas, M., Hoffmann, W., Piotrowski, D., Groppi, V., Allaman, G., Ogier, R., Bertrand, S., Bertrand, D. & Arneric, S. (2005). A novel positive allosteric modulator of the alpha7 neuronal nicotinic acetylcholine receptor: in vitro and in vivo characterization. *Journal of Neuroscience*, Vol.25, No.17, pp. 4396-4405, ISSN 1529-2401

Jacob, M.H., Lindstrom, J.M., & Berg, D.K. (1986). Surface and intracellular distribution of a putative neuronal nicotinic acetylcholine receptor. *The journal of cell biology*, Vol. 103, No. 1, (July 1986), pp. 205-214, ISSN 0021-9525

Kim H., Flanagin B.A., Qin C., Macdonald R.L., & Stitzel J.A. (2003). The mouse Chrna4 A529T polymorphism alters the ratio of high to low affinity α4β2 nAChRs. *Neuropharmacology* Vol.45, No.3 (September 2003), pp. 345–354, ISSN 0028-3908.

Langley, J. (1905). On the reaction of cells and of nerve-endings to certain poisons, chiefly as regards the reaction of striated muscle to nicotine and to curari.. *The Journal of Physiology*, Vol.33, No.4-5 (December 1905), pp. 374-413, ISSN 1469- 7793

Lindstrom, J., Anand, R., Gerzanich, V., Peng, X., Wang, F. and Wells, G. (1996). Structure and function of neuronal nicotinic acetylcholine receptors. *Progress in Brain Research*, Vol.109, No.1 (1996), pp. 125-137, ISSN 0079-6123

Ludwig J., Höffle-Maas A., Samochocki M., Luttmann E., Albuquerque E., Fels G., & Maelicke A. (2010). Localization by site-directed mutagenesis of a galantamine binding site on α7 nicotinic acetylcholine receptor extracellular domain. *Journal of Receptors and Signal Transduction*, Vol.30, No.6, (December 2006), pp. 469-483, ISSN 1532-4281

Luttmann, E., Ludwig, J., Höffle-Maas, A., Samochocki, M., Maelicke, A., & Fels G. (2009). Structural model for the binding sites of allosterically potentiating ligands on nicotinic acetylcholine receptors. *ChemMedChem*, Vol.4, No.11, (November 2009), pp. 1874-1882, ISSN 1860-7187

Macklin, K., Maus, A., Pereira, E., Albuquerque, E., & Conti-Fine, B. (1998). Human vascular endothelial cells express functional nicotinic acetylcholine receptors. *Journal of Pharmacology and Experimental Therapeutics*, Vol.287, No.1, (October 1998), pp. 435-439. ISSN 1521-0103

Maelicke, A., & Albuquerque, E. (2000). Allosteric modulation of nicotinic acetylcholine receptors as a treatment strategy for Alzheimer's disease. *European Journal of Pharmacology*, Vol.393, No.1-3, pp.165-170, ISSN 0014-2999

Maelicke, A., Samochocki, M., Jostock, R., Fehrenbacher, A., Ludwig, J., Albuquerque E., & Zerlin, M. (2001) Allosteric sensitization of nicotinic receptors by galantamine, a new treatment strategy for Alzheimer's disease. *Biological Psychiatry*, Vol.49, No.3, (February 2001), pp. 279-288, ISSN 0006-3223

Marks, M.J., Whiteaker, P., Calcaterra, J., Stitzel, J.A., Bullock, A.E., Grady, S.R. et al. (1999). Two pharmacologically distinct components of nicotinic receptor-mediated rubidium efflux in mouse brain require the β2 subunit. *The Journal of Pharmacology and Experimental Therapeutics* Vol.289, No.2 (May 1999), pp. 1090–1103, ISSN 0022-3565.

Maus, A., Pereira, E., Karachunski, P., Horton, R., Navaneetham, D., Macklin, K., Cortes, W., Albuquerque, E., & Conti-Fine, B. (1998). Human and rodent bronchial epithelial cells express functional nicotinic acetylcholine receptors. *Molecular Pharmacology*, Vol.54, No.5, (November 1998), pp. 779-788. ISSN 0026-895X

Mazzaferro, S., Benallegue, N., Carbone, A., Gasparri, F., Vijayan, R., Biggin, P.C., Moroni, M., & Bermudez, I. (2011). An additional ACh binding site at the α4/α4 interface of the (α4β2)$_2$α4 nicotinic receptor influences agonist sensitivity. *Journal of Biological Chemistry*, Vol.286, No.35 (September 2011), pp. 31043-31054, ISSN 0021-9258.

Miwa, J., Stevens, T., King, S., Caldarone, B., Ibanez-Tallon, I., Xiao, C., Fitzsimonds, R., Pavlides, C., Lester, H., Picciotto, M., & Heintz, N. (2006) The prototoxin lynx1 acts on nicotinic acetylcholine receptors to balance neuronal activity and survival in vivo. *Neuron*, Vol.51, No.5, (September 2006), pp. 587-600. ISSN 0896-6273

Moroni, M, Zwart, R., Sher, E., Cassels, B.K., & Bermudez, I. (2006). α4β2 nicotinic receptors with high and low acetylcholine sensitivity: pharmacology, stoichiometry, and sensitivity to long-term exposure to nicotine. *Molecular Pharmacology* Vol.70 No.2 (August 2006), pp. 755–768, ISSN 0026-895X

Nelson, M.E., Kuryatov, A., Choi, C., Zhou, Y., & Lindstrom, J. (2003). Alternate stoichiometries of α4β2 nicotinic acetylcholine receptors. *Molecular Pharmacology*, Vol.63, No.2 (February 2003), pp. 332-341, ISSN 0026-895X

Sambrook, J., & Russell, D. (2000). *Molecular Cloning: A Laboratory Manual, 3 Vol.* 3rd edition, Cold Spring Harbor Laboratory, ISBN 0879695773, New York

Schröder, B., Reinhardt-Maelicke, S., Schrattenholz, A., McLane, K., Kretschmer, A., Conti-Tronconi, B., & Maelicke, A. (1994). Monoclonal antibodies FK1 and WF6 define two neighboring ligand binding sites on Torpedo acetylcholine receptor alpha-polypeptide. *Journal of Biological Chemistry*, Vol.269, No.14 (April 1994), pp. 10407-10416, ISSN 0021-9258

Smit, A., Syed, N., Schaap, D., van Minnen, J., Klumperman, J., Kits, K., Lodder, H., van der Schors, R., van Elk, R., Sorgedrager, B., Brejc, K., Sixma, T., & Geraerts, W. (2001) A glia-derived acetylcholine-binding protein that modulates synaptic transmission. Nature, Vol.411, No.6835, (May 2001), pp. 261-268, ISSN 0028-0836

Storch, A., Schrattenholz, A., Cooper, J., Abdel Ghani, E., Gutbrod, O., Weber K., Reinhardt, S., Lobron, C., Hermsen, B., Soskiç, V., Pereira, E., Albuquerque, E., Methfessel, C., & Maelicke, A. (1995) Physostigmine, galanthamine and codeine act as 'noncompetitive nicotinic receptor agonists' on clonal rat pheochromocytoma cells.

European Journal of Pharmacology, Vol.290, No.3, (August 1995), pp. 207-219, ISSN 0014-2999

Takacs, Z., Wilhelmsen, K., & Sorota, S. (2001). Snake a-neurotoxin binding site on the Egyptian cobra (*Naja haje*) nicotinic acetylcholine receptor is conserved. *Mol. Biol. Evol.* Vol.18, No.9, pp.1800–1809, ISSN: 0737-4038

Tapia, L., Kuryatov, A., & Lindstrom, J. (2007). Ca^{2+} permeability of the $(\alpha4)_3(\beta2)_2$ stoichiometry exceeds that of $(\alpha4)_2(\beta2)_3$ human acetylcholine receptors. *Molecular Pharmacology*, Vol.71, No.3, (March 2007), pp. 769-776, ISSN 0026-895X

Temburni, M., Blitzblau, R., & Jacob M. (2000). Receptor targeting and heterogeneity at interneuronal nicotinic cholinergic synapses in vivo. *Journal of Physiology*, Vol.525.1 (March 2000), pp. 21-29, ISSN 0022-3751

Thomsen, T., Kaden, B., Fischer, J., Bickel, U., Barz, H., Gusztony, G., Cervos-Navarro, J., & Kewitz, H. (1991). Inhibition of acetylcholinesterase activity in human brain tissue and erythrocytes by galanthamine, physostigmine and tacrine. *European Journal of Clinical Chemistry and Clinical Biochemistry*, Vol.29, No.8, (August 1991), pp. 487-492, ISSN 1437-4331

Tritto, T., Stitzel, J.A., Marks, M.J., Romm, E., & Collins, A.C. (2002). Variability in response to nicotine in the LSXSS RI strains: potential role of polymorphism in alpha4 and alpha6 nicotinic receptor genes. *Pharmacogenetics* Vol.12, No.3 (April 2002), pp. 197-208, ISSN 0960-314X.

Tong, M., Arora, K., White, M. M., & Nichols, R. A. (2011). Role of key aromatic residues in the ligand-binding domain of alpha7 nicotinic receptors in the agonist action of beta-amyloid. *Journal of Biological Chemistry*, Vol.286, No.31, (September 2011), pp. 34373-34381, ISSN 0021-9258

Unwin, N. (2005). Refined structure of the nicotinic acetylcholine receptor at 4A resolution. Journal of Molecular Biology, Vol.346, No.4, (March 2005), pp.967-989, ISSN 0022-2836

Williams, B., Temburni, M., Levey, M., Bertrand, S., Bertrand, D., & Jacob, M. (1998). The long internal loop of the $\alpha3$ subunit targets nAChR to subdomains within individual synapses on neurones in vivo. *Nature Neuroscience*, Vol. 1 (November 1998), pp. 557-562 ISSN, 1097-6256

Wonnacott, S. (1997). Presynaptic nicotinic ACH receptors. *Trends in Neurosciences*, Vol. 20, No.2 (February 1997), pp. 92-98, ISSN 0166-2236

Xu, J., Zhu, Y., & Heinemann, S. (2006). Identification of sequence motifs that target neuronal nicotinic receptors to dendrites and axons. *The journal of neuroscience*, Vol.26, No.38, (September 2006), pp. 9780- 9793, ISSN 0270-6474

Young, G., Zwart, R., Walker, A., Sher, E., & Millar, N. (2008). Potentiation of alpha7 nicotinic acetylcholine receptors via an allosteric transmembrane site. *Proceedings of the National Acadademy of Sciences USA*, Vol.105, No.38, (September 2008) pp. 14686-1491, ISSN 1091-6490

Zhou, Y., Nelson, M.E., Kuryatov, A., Choi, C., Cooper, J., & Lindstrom, J. (2003). Human $\alpha4$ $\beta2$ acetylcholine receptors formed from linked subunits. *The Journal of Neuroscience*, Vol.23, Nr.27 (October 2003), pp. 9004 -9015, ISSN 0270-6474.

Zouridakis, M., Zisimopoulou, P., Poulas, K., & Tzartos, S. (2009). Recent advances in understanding the structure of nicotinic acetylcholine receptors. *IUBMB Life*, Vol.61, No.4, (April 2009), pp. 407-423, ISSN 1521-6551

6

Protein Engineering in Structure-Function Studies of Viper's Venom Secreted Phospholipases A2

Toni Petan[1], Petra Prijatelj Žnidaršič[2] and Jože Pungerčar[1]

[1]*Department of Molecular and Biomedical Sciences, Jožef Stefan Institute, Ljubljana,*
[2]*Department of Chemistry and Biochemistry, Faculty of Chemistry and Chemical Technology, University of Ljubljana, Ljubljana,*
Slovenia

1. Introduction

Secreted phospholipases A$_2$ (sPLA$_2$s) constitute a large family of interfacial enzymes that hydrolyze the *sn*-2 ester bond of membrane glycerophospholipids, releasing free fatty acids and lysophospholipids (Murakami et al., 2011). They are abundant in snake venoms, frequently being their major toxic components, and display a variety of pharmacological effects, such as neurotoxicity, myotoxicity, anticoagulant activity, cardiotoxicity and haemolytic activity. The molecular mechanisms underlying these effects are still poorly understood; but they are most probably based on the existence of specific, high-affinity binding sites for toxic sPLA$_2$s on the surface of target cells in specific target tissues (Kini, 2003). Snakes have developed, through evolution, an arsenal of structurally similar molecules that target a specific tissue or function, in order to capture and digest their prey. Interestingly, very often a particular sPLA$_2$ molecule may display, despite its simple, globular and compact structure, several different toxic activities. Thus, it significantly expands the possible prey-damaging mechanisms of the venom, which also depend on the type of prey, site of injection of the toxin and the tissue involved. The remarkable variety of pharmacological effects exerted by sPLA$_2$ toxins is a consequence of several factors that have intrigued scientists working in the field, but have also greatly complicated the study of their actions. The complications include (1) the apparently indiscriminate enzymatic activities of sPLA$_2$ toxins on different cellular and non-cellular phospholipid membranes and aggregates; (2) the diverse effects of the products of the hydrolysis of sPLA$_2$ toxins, especially on membrane structural integrity and dynamics, thus affecting the major structural and functional features of the cell; and (3) the ever increasing diversity of sPLA$_2$ intra- and extracellular binding proteins discovered in mammalian tissues, which are involved in very different biological processes. Additionally, many of the actions of sPLA$_2$ toxins involve complex, multi-step molecular mechanisms, in which a specific combination of enzymatic activity and/or protein binding is probably essential for a particular step. Although sPLA$_2$s are structurally highly conserved proteins, it is clear that subtle evolutionary changes of residues on the surface of the molecule have empowered these enzymes with this wide range of toxic activities. Their ability to recognize specific molecular targets has been gradually optimized and thus interferes with a range of physiological processes (Kini & Chan, 1999). Snakes have even developed, through their evolution, catalytically inactive sPLA$_2$-homologues specialized in membrane damage that occurs independently of enzymatic activity (Lomonte et al., 2009). Interestingly, a range of structurally very similar sPLA$_2$ enzymes, as well as an enzymatically inactive sPLA$_2$ homologue, are also present in mammals. The mammalian sPLA$_2$ family consists of 10 or 11 enzymes (Lambeau & Gelb, 2008) that display different cell- and tissue-specific expression patterns. The proteins act with a broad range of enzymatic activities on a variety of cellular and non-cellular phospholipid membranes (Murakami et al., 2011). They bind with high affinity to various soluble and membrane protein targets, many of which were discovered using toxic sPLA$_2$s (Pungerčar & Križaj, 2007; Valentin & Lambeau, 2000). Furthermore, apart from their direct effects on membrane structure and function, the products of their catalysis are precursors of hundreds of bioactive lipid signalling molecules, such as the eicosanoids. The mammalian sPLA$_2$ enzymes display a similarly broad range of roles, mostly incompletely understood and often contradictory, in various physiological and pathophysiological processes, such as lipid digestion and homeostasis, innate immunity, inflammation, fertility, blood coagulation, asthma, atherosclerosis, autoimmune diseases and cancer (Lambeau & Gelb, 2008; Murakami et al., 2011). In this they are analogous to their venom counterparts, owing their functions to a combination of enzymatic activity, direct and indirect effects of the products of their hydrolysis, and specific interactions with molecular partners inside or outside the cell. The research on the action of exogenous snake venom sPLA$_2$ enzymes, which target particular physiological processes in their mammalian prey, has been providing important clues for deciphering the biological roles of the mammalian endogenous sPLA$_2$ enzymes as well (Rouault et al., 2006; Valentin & Lambeau, 2000).

The most potent sPLA$_2$ toxins display presynaptic (ß-)neurotoxicity by attacking the presynaptic site of neuromuscular junctions. The venom of the nose-horned viper, *Vipera ammodytes ammodytes*, contains three presynaptically neurotoxic sPLA$_2$s, ammodytoxins (Atxs) A, B and C, two non-toxic ammodytins (Atns), AtnI$_1$ and AtnI$_2$, and a myotoxic and catalytically inactive Ser49 sPLA$_2$ homologue, ammodytin L (AtnL). They are all group IIA sPLA$_2$s. The presynaptically acting (ß-neurotoxic) Atxs interfere specifically with the release of acetylcholine from motoneurons and cause irreversible blockade of neuromuscular transmission. The exact mechanism of their action is not yet fully understood, but it must include specific binding to receptor(s) on the presynaptic membrane and enzymatic activity (Montecucco et al., 2008; Pungerčar & Križaj, 2007). The binding to highly specific, and yet unknown, primary molecular targets of ß-neurotoxins (ß-ntxs) on the presynaptic membrane is most probably followed by entry of the toxin into the nerve cell (Logonder et al., 2009; Pražnikar et al., 2008; Rigoni et al., 2008). It has been proposed that different sPLA$_2$-toxins exploit different internalization routes (Pungerčar & Križaj, 2007). In the motoneuron, they may impair the cycling of synaptic vesicles by phospholipid hydrolysis and by binding to specific intracellular protein targets, like calmodulin (Kovačič et al., 2009, 2010; Šribar et al., 2001) and 14-3-3 proteins (Šribar et al., 2003b) in the cytosol, and R25 (Šribar et al., 2003a) in mitochondria. Although the role of enzymatic activity in ß-neurotoxicity of sPLA$_2$s is still somewhat controversial, accumulated results speak largely in favour of its being indispensable for full expression of the ß-neurotoxic effect (Montecucco et al., 2008; Pungerčar & Križaj, 2007; Rouault et al., 2006).

In spite of numerous attempts to identify the surface residues of sPLA$_2$s crucial for a particular pharmacological effect (i.e., the "ß-neurotoxic site" or the "anticoagulant site") – based initially on structural analysis, chemical modification and, later on, site-directed mutagenesis, the molecular basis of their toxicity has yet to be resolved (Kini, 2003; Križaj, 2011; Pungerčar & Križaj, 2007; Rouault et al., 2006). We have addressed this issue in studies based on protein engineering of the nosed-horned viper sPLA$_2$s. These have resulted in more than fifty mutants and chimeric sPLA$_2$ proteins that have been produced and characterized in terms of their biochemical and biological activities (most of them are shown in Table 1). The site-directed mutagenesis studies have provided answers to, or at least significantly improved, our knowledge concerning many important questions regarding the toxic and enzymatic activities of Atxs and other sPLA$_2$s, such as:

- Which regions of the molecule are important for the ß-neurotoxicity of Atxs and homologous toxic sPLA$_2$s?
- Which surface residues of Atxs are crucial for interaction with their high-affinity binding proteins? Is there a correlation between the affinity for binding to a particular protein target and the neurotoxic potency of Atxs, suggesting a role for the receptor protein in the process?
- Which residues are responsible for interfacial membrane binding of Atxs? How do the interfacial membrane and kinetic properties of Atxs influence their toxic action? How do the enzymatic properties of Atxs compare with those of the best-studied mammalian sPLA$_2$s?
- What is the importance of the loss in evolution of enzymatic activity in the case of the AtnL and other myotoxic sPLA$_2$-homologues?

The results obtained have contributed significantly to a better understanding of the molecular mechanisms of action of snake venom sPLA$_2$s and provided clues to the action of the homologous groups of mammalian sPLA$_2$s.

2. Search for the "neurotoxic site" and the role of enzymatic activity

The significant structural similarities of toxic and non-toxic sPLA$_2$s, which differ in their pharmacological actions, have enticed a large number of researchers hoping to find the "holy grail" of sPLA$_2$-toxin research – the toxic site. The site is presumed to comprise only a small number of crucial amino acid residues (Kini, 2003). To explain the wide range of pharmacological effects induced by snake venom sPLA$_2$s, Kini & Evans (1989) proposed a model comprising specific "target sites" present on the surface of particular cell types. The target sites are proposed to be recognized by complementary "pharmacological sites" on the toxin molecule, these being structurally distinct from, and independent of, the "catalytic site." Thus, high-affinity binding (at least in the nM range) of the toxin to specific target sites ensures that, upon entering the circulation, each toxin binds primarily to its proper target tissue. It is highly likely that the primary target, or acceptor, sites are proteins. This is because of the much lower affinity (mM–μM range) of sPLA$_2$s for binding to the abundant zwitterionic phospholipid surfaces (i.e., cell membranes) (Bezzine et al., 2002; Petan et al., 2005; Singer et al., 2002) and, following enzyme adsorption to the membrane surface, indiscriminate binding to and hydrolysis of phospholipid molecules at the catalytic site. Therefore, separate pharmacological sites on an sPLA$_2$ molecule recognizing different target binding sites should be the main structural determinants that differentiate their respective pharmacological actions, such as presynaptic or central neurotoxicity. However, according to the results of our mutagenesis structure-function studies of Atxs reviewed below, it is unlikely that there is a structurally distinct, single "presynaptic neurotoxic site" located in a specific part of the molecule, in contrast to the strict physical localization of the enzyme active site. Rather, different parts of the toxin molecule are likely to be involved in different stages of the complex multi-step process of neurotoxicity, all contributing to the final outcome. In this view, structurally different ß-ntxs may have different surface regions that bind to different (extra- and intracellular) targets, which are nevertheless involved in the same process, most probably the recycling of synaptic vesicles. However, they all share the nonspecific $sPLA_2$ activity, i.e., the ability to bind and hydrolyze different phospholipid molecules embedded in membranes of various compositions – an essential step in the complete, irreversible blockade of neuromuscular transmission. Therefore, at least in the case of ß-neurotoxic $sPLA_2$s, the use of the term "presynaptic neurotoxic site" appears unsuitable for describing the multiple regions distributed on the surface of the $sPLA_2$ molecule (Prijatelj et al., 2008).

Given the multi-step, and as yet incompletely known, molecular events leading to presynaptic neurotoxicity of $sPLA_2$s, a simple correlation between their *in vitro* enzymatic activity and their lethal potency would not be expected (Rosenberg, 1997). Indeed, there are numerous examples of $sPLA_2$ ß-ntxs with significantly different enzymatic properties, which are, however, not reflected in differences in toxicity; in fact even the most potent $sPLA_2$ ß-ntxs are weak enzymes (Petan et al., 2005; Pražnikar et al., 2008; Prijatelj et al., 2006b, 2008; Rosenberg, 1997). Nevertheless, $sPLA_2$ enzymatic activity is necessary for full expression of the ß-neurotoxic effect (Montecucco et al., 2008; Pungerčar & Križaj, 2007). Its role in the process is most probably obscured by the numerous factors affecting both $sPLA_2$ activity and neurotoxicity, especially the, as yet unknown, (sub)cellular location, accessibility, composition and physical properties of the target membrane. The enzymatic action of $sPLA_2$s could lead to structural and functional destruction of cell membranes and organelles, like mitochondria or synaptic vesicles (Pražnikar et al., 2008, 2009; Pungerčar & Križaj, 2007; Rigoni et al., 2008), since the products of phospholipid hydrolysis are disruptive to many physiological processes by impairing the function of peripheral and integral membrane proteins and promoting membrane dysfunction by altering membrane asymmetry, curvature and fusogenicity (Montecucco et al., 2008; Paoli et al., 2009; Rigoni et al., 2005). The apparent lack of correlation between *in vitro* enzymatic activity and lethal potency of Atxs or other neurotoxic $sPLA_2$s (Montecucco et al., 2008; Petan et al., 2005; Pungerčar & Križaj, 2007) can be explained by the strict localization of the $sPLA_2$ activity to particular target membrane(s) due to binding to highly specific extra- and intracellular protein acceptors (Paoli et al., 2009; Petan et al., 2005; Pungerčar & Križaj, 2007). Our studies investigating the interfacial binding and kinetic properties of toxic $sPLA_2$s and their mutants have provided important clues to understanding the role of enzymatic activity in the process of presynaptic neurotoxicity of $sPLA_2$s. As described in detail below, despite their potent neurotoxic activity, Atxs are quite effective in hydrolysing pure phosphatidylcholine (PC) vesicles as well as PC-rich plasma membranes of mammalian cells, similarly to the most active mammalian group V and X $sPLA_2$ enzymes (Petan et al., 2005, 2007; Pražnikar et al., 2008). We have also shown that, when tightly bound to the membrane surface, the Ca^{2+} requirements of Atxs are in the micromolar range (Petan et al., 2005), opening up the possibility that such neurotoxins are also catalytically active in the subcellular compartments where Ca^{2+} concentrations are low (Kovačič et al., 2009; Petan et al., 2005). Moreover, Atxs are rapidly internalized in motoneuronal cells and are, surprisingly, translocated to the cytosol, where they specifically bind calmodulin (CaM) and 14-3-3 proteins, strongly opposing the dogma of the exclusively extracellular action of not only sPLA₂-neurotoxins, but also of sPLA₂s in general (Pražnikar et al., 2008). In agreement with these findings, we have recently shown that high-affinity binding to the cytosolic Ca^{2+}-sensor molecule CaM leads to structural stabilization (increased resistance to the reducing environment of the cytosol) and a significant augmentation of the enzymatic activity of Atxs and, intriguingly, also of the mammalian group V and X sPLA₂s (Kovačič et al., 2009, 2010). These findings strongly support the possibility of augmentation of Atx enzymatic activity by CaM in the cytosol during the process of ß-neurotoxicity. They also point to a new mechanism of modulating the enzymatic activity of mammalian group V and X sPLA₂s or some other non-toxic endogenous sPLA₂ (Kovačič et al., 2010).

3. Structural determinants of presynaptic neurotoxicity of sPLA₂s

The subtlety of the structure-function relationship of sPLA₂ neurotoxins is obvious on examination of the primary structures and toxicities of Atxs. The three sPLA₂ toxins each consist of 122 amino acid residues and differ at only five positions (Križaj, 2011). AtxC may be considered as a natural double mutant (F124I/K128E) and AtxB as a triple mutant (Y115H/R118M/N119Y) of AtxA. Nevertheless, their lethal potencies in mice differ considerably. AtxA is the most lethal; and its protein isoforms, AtxC and AtxB, are 17- and 28-fold less potent, respectively (Thouin et al., 1982). The crystal structures of recombinant AtxA (PDB code 3G8G) and natural AtxC (PDB code 3G8H) demonstrate the absence of significant structural differences between the two toxins (Saul et al., 2010). There is only a minor conformational difference at positions 127 and 128 in the C-terminal region, caused by the charge-reversal substitution of Lys128 for Glu, which does not significantly influence the toxicity (Saul et al., 2010). An illustrative example of the subtle structure-function relationships of sPLA₂s is the conversion, by a single mutation (F22Y), of the gene for bovine pancreatic group IB sPLA₂ to a gene encoding a molecule able to compete with crotoxin, a ß-neurotoxic sPLA₂ from the South American rattlesnake, *Crotalus durissus terrificus*, for binding to its 45-kDa neuronal-binding protein. This led the authors to suggest the conversion of the non-toxic pancreatic sPLA₂ to a neurotoxic molecule (Tzeng et al., 1995).

By substituting several basic residues in the C-terminal region (AtxA[NNTETE] mutant: AtxA-K108N/K111N/K127T/K128E/E129T/K132E) and in the ß-structure region (AtxA[SSL] mutant: AtxA-K74S/H76S/R77L) with acidic and non-ionic residues, we have shown, contrary to previous beliefs, that the basic character of Atxs, and probably of other ß-neurotoxic sPLA₂s, is not obligatory for presynaptic toxicity (Ivanovski et al., 2004; Prijatelj et al., 2000; Table 1). According to our earlier structure-function analyses, the more than one order of magnitude lower toxicity of AtxC than that of AtxA is a consequence of the substitution of the aromatic Phe124 by Ile (Pungerčar et al., 1999). Furthermore, in accordance with the three substitutions responsible for the difference in toxicities of AtxA and AtxB, several other C-terminal residues of AtxA, namely the Tyr115/Ile116/Arg118/Asn119 (YIRN) cluster, were shown to be important for the neurotoxicity of Atxs (Ivanovski et al., 2000). Thus, the lethal potency of the AtxA-Y115K/I116K/R118M/N119L (AtxA[KKML]) mutant was 290-fold lower than that of AtxA (Ivanovski et al., 2000 and Table 1).

sPLA$_2$	LD$_{50}$ (µg/kg)	IC$_{50}$ (nM)			References
		CaM	R25	R180	
AtxA	21	6 ± 2	10 ± 3	16 ± 3	a, b
AtxB	580	23 ± 4	n. d.	n. d.	a, c
AtxC	360	21 ± 3	50	155	a, c, d, e
12-AtxA	280	72 ± 15	5 ± 2	> 10^4	f
I-AtxA	500	250 ± 60	3.4 ± 0.3	> 10^4	f
P-AtxA	420	380 ± 85	3.5 ± 0.5	> 10^4	f
AtnI$_2$/AtxA^{K108N}	> 10^4	1300 ± 200	20 ± 6	490 ± 100	g
AtnI$_2$^{N24F}/AtxA^{K108N}	> 5000	1700 ± 300	24 ± 6	850 ± 200	g
AtnI$_2$	> 10^4	> 10^4	> 10^4	610 ± 100	g
AtxA^{NNTETE}	660	27 ± 5	16	100	c, d
AtxA$^{K108N/K111N}$	67	17 ± 4	38	68	c, d, h
AtxA^{K127T}	35	20 ± 3	22	300	c, d
AtxA^{K128E}	45	14 ± 3	n. d.	n. d.	c, h
AtxA^{F24A}	90	7.0 ± 0.9	15 ± 1	17 ± 2	b
AtxA^{F24N}	2800	5.6 ± 0.7	14 ± 2	26 ± 1	b
AtxA^{F24S}	380	7.4 ± 0.2	14 ± 1	16 ± 2	b
AtxA^{F24W}	175	13.6 ± 0.3	37 ± 5	26 ± 3	b
AtxA^{F24Y}	330	9.2 ± 0.9	14 ± 3	19 ± 4	b
AtxA^{V31W}	135	n. d.	n. d.	n. d.	i
AtxA^{R72E}	84	71 ± 3	14 ± 1	78 ± 7	j
AtxA^{R72I}	32	17 ± 2	18.0 ± 0.4	28 ± 3	j
AtxA^{R72K}	50	46 ± 2	11.7 ± 0.4	83 ± 9	j
AtxA^{R72S}	55	24 ± 2	16 ± 1	35 ± 1	j
AtxA^{SSL}	276	18 ± 1	18 ± 1	107 ± 8	j
AtxA^{K86A}	24	8.1 ± 0.3	15 ± 1	20 ± 3	j
AtxA^{K86E}	32	7 ± 1	18 ± 1	42 ± 4	j
AtxA^{K86G}	34	7.5 ± 0.2	12.1 ± 0.1	65 ± 4	j
AtxA^{K86R}	31	8.8 ± 0.5	16.3 ± 0.5	36 ± 1	j
AtxA^{KKML}	~ 6000	50 ± 9	86	257	c, k
AtxA^{KK}	~ 5000	21 ± 3	380	118	c, k
AtxA/DPLA$_2$^{YIRN}	45	27 ± 5	180 ± 25	100 ± 16	l
AtxA^{KEW}/DPLA$_2$^{YIRN}	910	43 ± 12	110 ± 15	22 ± 4	l
AtxA^{KE}/DPLA$_2$^{YIRN}	790	14 ± 4	87 ± 9	24 ± 4	l
AtxA^{W}/DPLA$_2$^{YIRN}	107	28 ± 7	78 ± 12	19 ± 5	l
AtxA/DPLA$_2$	2600	110 ± 10	45%#	280 ± 19	l
DPLA$_2$^{YIRN}	~ 17000	43 ± 14	200 ± 30	120 ± 21	c, l
DPLA$_2$	3100	300 ± 36	75%#	300 ± 45	c, l
AtnL	> 10000	n. d.	n. d.	n. d.	m
AtnL^{YVGD}	> 7000	n. d.	n. d.	n. d.	m
AtnL^{YWGD}	2200	n. d.	n. d.	n. d.	m

Table 1. Lethal potency and protein-binding affinity of Atxs, Atns, DPLA$_2$ and their mutants. IC$_{50}$ values (the concentration of competitor sPLA$_2$ required to reduce the binding of 10 nM ^{125}I-AtxC by 50%) were determined from competition binding experiments for binding to calmodulin (CaM), the mitochondrial receptor R25 and the neuronal M-type sPLA$_2$ receptor, R180. #The recombinant toxin did not completely inhibit the binding of ^{125}I-AtxC. [a]Thouin et al., 1982; [b]Petan et al., 2002; [c]Prijatelj et al., 2003; [d]Prijatelj et al., 2000; [e]Čopič et al., 1999; [f]Prijatelj et al., 2006b; [g]Prijatelj et al., 2002; [h]Pungerčar et al., 1999; [i]Petan et al., 2005; [j]Ivanovski et al., 2004; [k]Ivanovski et al., 2000; [l]Prijatelj et al., 2008; [m]Petan et al., 2007.

The KKML cluster is present in the weakly neurotoxic DPLA$_2$ from the venom of Russell's viper, *Daboia (Vipera) russelii russelii*, which shares a high level of amino acid identity (82%) with AtxA. However, the latter is almost 150-fold more toxic in mice (Prijatelj et al., 2003). To our great surprise, the introduction of the YIRN cluster into DPLA$_2$ did not increase its toxicity; on the contrary, the DPLA$_2$[YIRN] mutant was more than five times less toxic than DPLA$_2$ (Prijatelj et al., 2003). Additionally, our study on the importance of the N-terminal residue Phe24, in which it was replaced by other aromatic (tyr or trp), polar uncharged (ser or asn) or hydrophobic (ala) residues, suggested that Phe24 is also involved in the neurotoxicity of Atxs, apparently at a stage not involving enzymatic activity or interactions with the high-affinity binding proteins R25, R180 and CaM (Petan et al., 2002). The aromatic Phe24 was chosen for this study on the basis of several interesting characteristics. It is located in a region immediately preceding the Ca^{2+}-binding loop, but it is spatially close to the important Phe124. It is important for membrane binding as part of the interfacial binding surface (IBS, see below) of the enzyme, and it is replaced by ser in the weakly neurotoxic DPLA$_2$. These facts prompted us to propose that a particular combination of both C-terminal and N-terminal residues must be involved in ß-neurotoxicity. In order to identify the N-terminal residues that supplement the role of the YIRN cluster in the high neurotoxic potency of AtxA, we selectively mutated some of the remaining residues that differentiate DPLA$_2$ from AtxA. First, we introduced the N-terminal half of AtxA into DPLA$_2$ by preparing the chimeric AtxA/DPLA$_2$ protein. Its lethal potency was relatively low, in the range of those of DPLA$_2$ and the AtxA[KKML] mutant (Table 1 and Ivanovski et al., 2000), confirming that it is primarily the presence of the KKML cluster in the C-terminus of the chimera that has a strong negative influence on toxicity. Secondly, by substituting the KKML cluster in DPLA$_2$ with the YIRN cluster of AtxA, we produced a chimeric mutant (AtxA/DPLA$_2$[YIRN]) that is 58-fold higher in lethal potency than is AtxA/DPLA$_2$, reaching a level of toxicity similar to that of the highly neurotoxic AtxA. Thus, only in combination with the N-terminal part of AtxA is the presence of the YIRN cluster sufficient for the high neurotoxic potency of AtxA and AtxA/DPLA$_2$[YIRN]. This allowed us to exclude the importance of the additional eleven C-terminal residues present in AtxA and absent in DPLA$_2$ – Thr70, His76, Glu78, Gly85, Arg100, Asn114, Ser130, Glu131 – and was in accordance with the findings of our early mutagenesis study on the three C-terminal lysines at positions 108, 111 and 128 (Pungerčar et al., 1999) (Figure 1). These results clearly confirmed our hypothesis that a particular combination of C-terminal residues, especially those in the region 115-119, i.e., the YIRN cluster, and certain N-terminal residues are necessary for the potent ß-neurotoxicity of Atxs.

Our next objective was to determine the contribution to the high neurotoxic potency of AtxA of the remaining nine N-terminal residues that differentiate it from DPLA$_2$ (Met7, Gly11, Asn17, Pro18, Leu19, Thr20, Phe24, Val31 and Ser67). As described above, the importance of the aromatic Phe24 had already been established (Petan et al., 2002). In order to assess the importance of the remaining N-terminal residues, we first substituted Met7, Gly11 and Val31 in the highly toxic chimera AtxA/DPLA$_2$[YIRN] by the corresponding residues present in DPLA$_2$ – lys, glu and trp, respectively. The mutant protein AtxA[KEW]/DPLA$_2$[YIRN] displayed a 20-fold lower lethal potency than the AtxA/DPLA$_2$[YIRN] chimera, suggesting involvement of the group of residues at positions 7, 11 and 31 in AtxA neurotoxicity. The lethality of the partial mutants, AtxA[KE]/DPLA$_2$[YIRN] and AtxA[W]/DPLA$_2$[YIRN] (Table 1), revealed that the contribution of the pair of residues in the N-terminal helix, Met7 and Gly11, to the

```
                    1        10        20        30        40        50        60
AtxA                SLLEFGMMILG-ETGKNPLTSYSFYGCYCGVGGKGTPKDATDRCCFVHDCCYGNLP--D-C-----S
AtxB                ...........-.....................................--.-.-----.
AtxC                ...........-.....................................--.-.-----.
AtxAKKML            ...........-.....................................--.-.-----.
AtxA/DPLA2YIRN      ...........-.....................................--.-.-----.
AtxAKEW/DPLA2YIRN   ......K...E-....................W................--.-.-----.
AtxA/DPLA2          ...........-.....................................--.-.-----.
DPLA2YIRN           ......K...E-....LAIP...S......W..................--.-.-----N
DPLA2               ......K...E-....LAIP...S......W..................--.-.-----N
AtnL                .VI...K..QE-..D............H..L.N..K.............S...AK.S--.-.-----.
AtnLYVGD            .VI...K..QE-..D................V....K................AK.S--.-.-----.
AtnLYWGD            .VI...K..QE-..D................W....K................AK.S--.-.-----.
AtnI2               N.YQ..N..FK-M.K.SA.L...N......W....K.Q.............RVN--G-.-----D
hGIIA               N.VN.HR..KL-T...EAAL..G....H.....R.S........VT.....KR.EKRG-.-----G
hGV                 G..DLKS..EK-V....A..N.G.......W..R.....G..W..WA..H...R.EEKG-.-----N
hGX                 GI..LAGTVGC-V-.PRTPIA.MK...F..L..H.Q.R..I.W..HG.....TRAEEAG-.-----.
```

```
                    70        80        90        100       110       120       130
AtxA                PKTDRYKYHRENGAIVCGK-GTSCENRICECDRAAAICFRKNLKTYNYIYRNYPD-FLCKKESEKC
AtxB                ..................-............................H..MY...-........
AtxC                ..................-.............................-I...E.....
AtxAKKML            ..................-...........................KK.ML...-........
AtxA/DPLA2YIRN      ..S.....K.V.....E.-..............K.......Q..N..S.........-....G.-L..
AtxAKEW/DPLA2YIRN   ..S.....K.V.....E.-..............K.......Q..N..S.........-....G.-L..
AtxA/DPLA2          ..S.....K.V.....E.-..............K.......Q..N..SKK.ML...-....G.-L..
DPLA2YIRN           ..S.....K.V.....E.-..............K.......Q..N..S.........-....G.-L..
DPLA2               ..S.....K.V.....E.-..............K.......Q..N..SKK.ML...-....G.-L..
AtnL                ...N..E...........S-S.P.KKQ.............E......KK.KV.LR-.K..GV....
AtnLYVGD            ...N..E...........S-S.P.KKQ.............E......KK.KV.LR-.K..GV....
AtnLYWGD            ...N..E...........S-S.P.KKQ.............E......KK.KV.LR-.K..GV....
AtnI2               ..LSI.S.SF....D....G-DDP.LRAV.....V.....GE..N..DKK.K...S-SH.TET-.Q.
hGIIA               T.FLS..FSNSGSR.T.A.-QD..RSQL....K...T..AR.KT...KK.QY.SN-KH.RGSTPR.
hGV                 IR.QS...RFAW.VVT.EP-.PF.HVNL.A...KLVY.LKR..RS..PQ.QYF.N-I..S
hGX                 ...E..SWQCV.QSVL..PAENK.QELL.K..QEI.N.LAQT--E..LK.LF..Q-...EPD.P..D
```

Fig. 1. Amino acid alignment of snake venom group IIA sPLA₂s, including the sPLA₂-homologue ammodytin L (AtnL), some of their mutants and the human group IIA, V and X sPLA₂s. The residues comprising the putative IBS of Atxs are presented in bold type, while those most important for the neurotoxicity of ammodytoxins (Atxs) are underlined. The weakly neurotoxic sPLA₂ from Russell's viper, *Daboia r. russelii*, DPLA₂, differs from AtxA in only 22 residues (82% identity). AtnL, the enzymatically inactive but myotoxic Ser49 structural homologue of Atxs, displays 74% amino acid identity with AtxA. The neutral ammodytin I_2 (AtnI$_2$) is a non-toxic homologue of Atxs from the same venom with 58% amino acid identity with AtxA. Atxs also display a relatively high degree of identity (48%) with the human groups IIA (hGIIA), V (hGV) sPLA$_2$s, and X (hGX) sPLA$_2$ (41%). The common sPLA$_2$ numbering of residues was used (Renetseder et al., 1985). Gaps, represented by dashes, were used to align the homologous sPLA$_2$s according to conserved residues. Identical amino acid residues are shown by dots. Amino acid single-letter symbols are shown in Table 1 in the chapter by Figurski et al.

neurotoxic potency of AtxA is substantially higher than that of Val31 – in accordance with the negligible effect of the V31W mutation on the lethality of AtxA, despite its outstanding effect on enzymatic activity (Petan et al., 2005). Interestingly, the bulky Trp at position 31 of the AtxA^{V31W} and AtxA^W/DPLA$_2$^{YIRN} mutants had no significant impact on neurotoxicity, despite being spatially very close to Phe24 and also to the YIRN cluster. The collective contribution of the N-terminal residues Met7, Gly11 and Phe24 to the neurotoxicity of AtxA is very significant and similar to that of the YIRN cluster in the C-terminus. For example, the substitution of Phe24 with ser in AtxA caused an approximately 19-fold decrease in neurotoxicity (Petan et al., 2002), which is similar to the reduction seen after adding the N-terminal region of AtxA^{KEW}, in which the residues Met7, Gly11 and Val31 were substituted with lys, glu and trp, respectively, to DPLA$_2$^{YIRN}, resulting in the AtxA^{KEW}/DPLA$_2$^{YIRN} chimeric mutant. Therefore, it is highly likely that the remaining N-terminal residues differentiating AtxA and DPLA$_2$, i.e., the Asn17/Pro18/Leu19/Thr20 cluster and Ser67, are not greatly involved in neurotoxicity.

Our structure-function studies of ß-neurotoxic sPLA$_2$s (Ivanovski et al., 2000, 2004; Petan et al., 2002, 2005; Prijatelj et al., 2000, 2002, 2003, 2006b, 2008; Pungerčar et al., 1999) clearly show that different parts of the toxin molecule have separate roles in the distinct steps of the complex mechanism of presynaptic neurotoxicity (Pungerčar & Križaj, 2007). Most significantly, by selectively mutating parts of the DPLA$_2$ molecule, we were able to map the residues that separate the weakly ß-neurotoxic sPLA$_2$ from the 150-fold more potent AtxA (Prijatelj et al., 2008). In summary, the most important structural features responsible for the high neurotoxic potency of Atxs (Figure 2, A) are: the "upper" part of the molecule concentrated around the C-terminal region 115-119 (the YIRN cluster) and including the spatially close aromatic Phe124 and Phe24, and the N-terminal helix region with the Met7/Gly11 pair in the "lower right" part of the molecule.

4. Enzymatic activity of Atxs: factors influencing interfacial binding and hydrolysis of phospholipid membranes

Secreted PLA$_2$s are prototypical interfacial enzymes (Berg et al., 2001). In order to gain access to their phospholipid substrate, sPLA$_2$s first have to bind at the lipid/water membrane interface by their interfacial binding surface (IBS), a group of residues located on a relatively flat exposed region surrounding the entrance to the active site pocket (Lin et al., 1998; Ramirez & Jain, 1991). Only then, after the enzyme is bound to the surface, can binding of a single phospholipid molecule and its hydrolysis occur in the active site. The catalytic turnover cycle of sPLA$_2$s includes a highly conserved His48–Asp99 dyad and an activated water molecule that acts as a nucleophile during hydrolysis of the sn-2 ester bond of the phospholipid (Scott et al., 1990). The resulting tetrahedral intermediate is stabilized by the

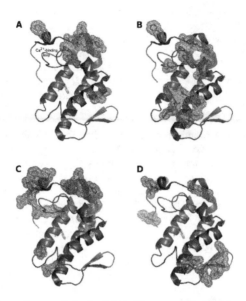

Fig. 2. Surface residues of ammodytoxin A (AtxA) responsible for A) presynaptic neurotoxicity, B) interfacial membrane binding, C) binding to calmodulin and D) binding to factor Xa. The molecule is oriented with the interfacial binding surface (IBS) and the N-terminal residues facing the viewer, while the C-terminal region is located in the upper-left corner of the molecule; and the ß-structure, in the lower-right corner. A) The conserved sPLA$_2$ helical structures are labelled with alphabet letters (red) (see section 4.1 for details; Saul et al., 2010). Residues important for neurotoxicity are presented in orange mesh surface and extend from the C-terminal region 115–119 (the YIRN cluster), including the spatially close aromatic Phe124 and Phe24, to the N-terminal helix region with the Met7/Gly11 pair in the "lower right" part of the molecule. B) The IBS residues of AtxA are presented in blue mesh surface (Leu2, Leu3, Leu19, Thr20, Phe24, Val31, Ser67, Lys69, Thr70, Arg72, Arg118, Asn119 and Phe124) and surround the active site pocket with His48 and Asp99 (presented in green). C) The Atx–CaM interaction surface comprises most of the C-terminal residues of AtxA in the region 108–131 and several basic residues in α-helices C and D (red mesh surface representation; more details in Kovačič et al., 2010). D) The anticoagulant activity of AtxA is a consequence of its binding to factor Xa, which involves mostly basic residues (magenta mesh surface): Arg118, Lys127, Lys128 and Lys132 in the C-terminal end and Arg72, Lys74, His74 and Arg77 in the ß-structure region (Prijatelj et al., 2006a).

Ca^{2+} cofactor, which is coordinatively bound by three main-chain carbonyl oxygen atoms of residues in the conserved Ca^{2+}-binding loop of the enzyme (Tyr28, Gly30 and Gly32) by two carboxylate oxygen atoms of Asp49, and by two oxygen atoms from the phospholipid substrate (Scott et al., 1990). Calcium ion is required for the initial binding of a phospholipid molecule to the active site of sPLA$_2$s and for the catalytic step, but it is not necessary for adsorption of sPLA$_2$s to the membrane (Yu et al., 1993). Given that the binding of sPLA$_2$s to a membrane surface is structurally and kinetically independent of the subsequent binding and catalytic steps at the active site (Berg et al., 2001), the term substrate specificity in the case of sPLA2s is a combination of two independent "specificities": (1) the affinity of the enzyme for binding to a membrane surface, governed by the interaction of the IBS and 20–40 phospholipids on the surface, and (2) the relative velocity of hydrolysis of different phospholipid species by the membrane-bound enzyme that obeys Michaelis-Menten kinetics (Berg et al., 2001; Lambeau & Gelb, 2008). The latter is determined by many factors influencing the interactions of the substrate molecule in the active site cleft, the rate of the catalytic reaction and the rate of release of the reaction products. In general, the active sites of sPLA2s display low specificity for different phospholipid head-groups and acyl chains (Singer et al., 2002). As a consequence, the physiological functions of some sPLA2s, e.g., the human group IIA (hGIIA), V (hGV) and X (hGX) enzymes, are significantly influenced or even determined by their different interfacial binding specificities and not by the specificity of their catalytic sites (Beers et al., 2003; Bezzine et al., 2000, 2002; Pan et al., 2002; Singer et al., 2002). Therefore, factors influencing interfacial binding of mammalian and toxic sPLA2s, such as the composition and physical properties of the membrane, the nature of the IBS of the enzyme, and the concentration of phospholipids that are accessible to the sPLA2 (Mounier et al., 2004), are crucial determinants of sPLA2 biological activity.

The IBS of most sPLA2s comprises a ring of conserved hydrophobic residues, whereas, around them and on the edges of the IBS, some polar, basic and acidic residues are present (Petan et al., 2005; Snitko et al., 1997). Given that most sPLA2s have high activities on anionic membrane surfaces, it was long thought that electrostatic forces between cationic residues of the enzyme and anionic membrane phospholipids are crucial for the membrane binding and activity of sPLA2s (Scott et al., 1994). However, a number of studies have shown that, in fact, hydrophobic, aromatic and hydrogen-bonding interactions account for most of the binding energy, even to negatively charged membrane surfaces (Gelb et al., 1999; Ghomashchi et al., 1998; Lin et al., 1998). Nevertheless, it has been suggested that electrostatic interactions are responsible for proper orientation and initial association of sPLA2s with both anionic and zwitterionic membranes. They may significantly modulate the dynamics of establishing strong hydrophobic interactions, when aliphatic and aromatic residues on the IBS penetrate the membrane surface, needed to form a stable sPLA2-membrane complex (Beers et al., 2003; Petan et al., 2005; Prijatelj et al., 2008; Stahelin and Cho, 2001).

4.1 Basic structural, interfacial kinetic and binding properties of Atxs

Atxs are highly basic proteins (AtxA has a pI of 10.2, net charge +6), with structural features typical of group IIA sPLA2 enzymes (Figure 2, A): an N-terminal α-helix A (residues 1–14), a short α-helix B (residues 16–22), a Ca^{2+}-binding loop (residues 25–35), a long α-helix C (residues 39–57), a loop preceding an antiparallel two-stranded ß-sheet (ß-structure; residues 75–78 and 81–84), a long α-helix D (antiparallel to helix C; residues 89–109) and a C-terminal extension (mostly disordered, with two short helical turns; residues 110–133). The active site cleft is buried at the end of a hydrophobic channel (formed by the residues Phe5, Gly6, Tyr22, Ser23, Cys29, Gly30, Ala102, Ala103, Phe106) leading to the highly conserved His48–Asp99 dyad located between the antiparallel α-helices C and D. The entrance to the catalytic site of all sPLA2s is surrounded by the IBS group of residues, forming a relatively flat surface on one side of the molecule that is responsible for membrane binding – the first and prerequisite step in sPLA2 catalytic action. The IBS of Atxs (Figure 2, B) is formed by Leu2, Leu3, Leu19, Thr20, Phe24, Val31, Ser67, Lys69, Thr70,

Arg72, Arg118, Asn119, and Phe124 (Petan et al., 2005). The slight structural flexibility observed for the exposed side chains of residues in the IBS (e.g., Phe24 in Figure 2, B) is in keeping with the role of these residues in supporting optimal interactions of the molecule with the dynamic structure of phospholipid aggregates (Saul et al., 2010).

According to the enzymatic activities and membrane-binding affinities of Atxs determined on different phospholipid vesicles, it is clear that these snake venom sPLA$_2$s are very effective in binding to and hydrolyzing different phospholipid membranes (Petan et al., 2005). This property exists despite the fact that they have evolved to be specific and potent neurotoxic molecules and despite the fact that they share striking structural and functional similarities with the mammalian (non-toxic) sPLA$_2$s. The enzymatic activities of Atxs on vesicles composed of anionic phosphatidylglycerol (PG), which is, in general, the best sPLA$_2$ substrate, are comparable to those displayed by the most potent mammalian sPLA$_2$, the pancreatic group IB sPLA$_2$, and 5-fold higher than those displayed by the mammalian group IIA sPLA$_2$s on these vesicles (Singer et al., 2002). Atxs also have particularly high activities on phosphatidylserine (PS) vesicles, the main anionic phospholipid in eukaryotic membranes, well above the activities of the group IB and IIA sPLA$_2$s that are the most active mammalian sPLA$_2$s on these vesicles. Most importantly, the activities of Atxs were high also on pure zwitterionic phosphatidylcholine (PC) vesicles, much higher than that of the highly homologous and cationic hGIIA enzyme, which cannot bind to the PC-rich plasma membranes of mammalian cells (Beers et al., 2003; Bezzine et al., 2000, 2002; Birts et al., 2009; Singer et al., 2002). The hGIIA sPLA$_2$ is well known for its preference for anionic phospholipid substrates and its negligible activity on PC-rich membrane surfaces (Beers et al., 2003; Bezzine et al., 2000, 2002). These properties strongly influence the physiological role of the hGIIA enzyme, for example, by enabling its high antibacterial concentrations in human tears without affecting the corneal epithelial cells (Birts et al., 2009). Although the activities of Atxs were lower than those displayed by group V and X sPLA$_2$s, which are by far the most potent among the mammalian sPLA$_2$s in hydrolyzing PC vesicles and releasing fatty acids from cell membranes (Bezzine et al., 2000, 2002; Singer et al., 2002), they were able to hydrolyze plasma membranes of different intact mammalian cells at a rate that correlated well with their specific activities on PC vesicles (Petan et al., 2005; Pražnikar et al., 2008). The high activity of Atxs on PC-rich vesicles is a consequence of the ability of these venom sPLA$_2$s to bind well to such membrane surfaces, with affinities comparable to those of mammalian group V and X sPLA$_2$s. Furthermore, unlike the neutral hGX and similarly to the highly cationic hGIIA enzyme, the presence of anionic phospholipids in the membrane surface greatly enhances the membrane-binding affinity of Atxs and consequently the rate of phospholipid hydrolysis (Bezzine et al., 2000, 2002; Petan et al., 2005, 2007). This property may have an important influence on both localization of the toxin to its target membrane and its enzymatic effectiveness *in vivo*. It is now clear that, when conditions of high affinity binding apply (i.e., binding of hGIIA or Atxs to anionic PG vesicles), sPLA$_2$s reach their half-maximal enzymatic activities at low micromolar concentrations of Ca^{2+} (Petan et al., 2005; Singer et al., 2002). The proposed cytosolic action of Atxs (Pražnikar et al., 2008), and most probably other structurally similar sPLA$_2$s, is strongly supported by several factors. They may enable the enzymatic activity of Atxs in a reducing environment containing insufficient (nanomolar) concentrations of calcium. In support of this idea are (1) the high degree of stability of AtxA under conditions resembling those in the cytosol of eukaryotic cells (Kovačič et al., 2009; Petrovič et al., 2004), (2) the additional and very significant structural stabilization and augmentation of sPLA$_2$ enzymatic activity by CaM (Kovačič et al., 2009, 2010), (3) the transient cytosolic microdomains of high local calcium concentrations (~100 μM) (Meldolesi et al., 2002) or calcium entry through the damaged plasmalemma due to sPLA$_2$ action prior to internalization (Montecucco et al., 2008), and (4) the presence of anionic phospholipids (PS) on the cytosolic face of the plasma membrane and internal cellular organelles (Okeley & Gelb, 2004). In conclusion, Atxs are effective enzymes that bind strongly to and hydrolyze rapidly both anionic and zwitterionic phospholipid aggregates, including mammalian plasma membranes, presenting a broad combination of properties characteristic of different mammalian sPLA$_2$s. The potential for the high enzymatic activity of Atxs appears to be at odds with their specific neurotoxic action. However, it is in line with the possible limitations imposed on their intracellular activity on a particular target membrane by the harsh conditions in the cytosol during the process of neuromuscular transmission blockade.

4.2 Role of different IBS residues in supporting interfacial binding and activity of Atxs

The presence of tryptophan on the IBS of different sPLA$_2$s is a well-known determinant of their ability to bind with high affinity and hydrolyze PC-rich membranes, crucially influencing their biological roles. Its role has been highlighted in the cases of the hGV and hGX enzymes (Bezzine et al., 2002; Han et al., 1999), the acidic sPLA$_2$ from *Naja naja atra* snake venom (Sumandea et al., 1999), a range of mutants of hGIIA (Beers et al., 2003) and the pancreatic group IB sPLA$_2$s (Lee et al., 1996). Despite the fact that Atxs do not contain a Trp residue, they display a relatively high activity on PC vesicles. Furthermore, the substitution of Val31 by Trp led to a dramatic 27-fold increase in the activity of AtxA (Petan et al., 2005), reaching a level of activity on PC-rich vesicles higher than those of hGV and hGX and in the range of the best-acting snake venom sPLA$_2$s, e.g., the cobra venom sPLA$_2$ (Sumandea et al., 1999). However, despite its very high affinity for PC-rich surfaces and its order of magnitude higher potency in releasing fatty acids from plasma membranes of intact HEK293 cells, C2C12 myocytes and motoneuronal NSC34 cells (Pražnikar et al., 2008), the AtxA-V31W mutant did not display a major change in neurotoxic potency *in vivo*. This again highlights the dependence of the toxicity of sPLA$_2$ ß-ntxs on a combination of toxin-acceptor interactions leading to localization of the toxin to the target membrane at which the enzymatic activity of AtxA is sufficient for the observed effects. In contrast, the same substitution, V31W, on the IBS of the enzymatically active quaternary mutant AtnL^{YVGD} of the myotoxic Atx homologue, AtnL (see below), caused a 100-fold increase in its activity on PC vesicles, as well as a significant increase in its toxicity *in vivo* and *in vitro* (Petan et al., 2007). Thus, in the case of the enzymatically active mutants of AtnL, the correlation between enzymatic activity on PC membranes and toxicity *in vivo* suggests that the mechanism of their toxicity differs from that used by Atxs and that it depends largely on their interfacial binding affinity and enzymatic activity on PC-rich target membranes, both of which are significantly affected by Trp31 (Petan et al., 2007).

The role of aromatic residues in the interfacial binding of sPLA$_2$s depends to a high degree on the nature of the residue itself, its position on the IBS and the orientation of its side-chain (Stahelin & Cho, 2001; Sumandea et al., 1999). This is evident from the fact that the substitution of Phe24 with Trp did not cause a substantial increase in enzymatic activity or interfacial binding affinity of AtxA (Petan et al., 2002, 2005) or hGIIA (Beers et al., 2003).

Thus, both Phe and Trp can have similar roles in interfacial binding, despite the differences in their interactions with the membrane: the highly amphiphilic Trp favours partitioning in the interfacial phospholipid head group region of the bilayer (Yau et al., 1998), while the aromatic Phe penetrates deeper into the hydrophobic core of the phospholipid acyl chains (Stahelin & Cho, 2001; Sumandea et al., 1999). In the case of AtxA and hGIIA, Trp31 is obviously in a much better position to influence interfacial binding than is Trp24. Additionally, AtxA and AtxA^{F24W} display higher activities and binding affinities than do the F24S, F24Y and F24N mutants on PC-rich vesicles containing anionic phospholipids. Although the structures of Trp and Phe differ significantly, both are obviously better suited to take advantage of the presence of anionic phospholipid in the interface than the polar Ser, Asn and even the aromatic Tyr. The importance of non-polar interactions in interfacial binding to negatively charged surfaces is also clear in the case of AtxB, AtxA^{KKML} and AtxA^{V31W}. These molecules already display very high activities on PC vesicles containing 10% PS (10% PS/PC vesicles), and they reach the level of maximal activity already on 30% PS/PC vesicles, indicating that the enzymes are fully bound to these vesicles (Petan et al., 2005). In general, we have observed the greatest increases in activity upon introduction of 10% anionic phospholipids (PS or PG) into PC vesicles for mutants that have numbers of hydrophobic and aromatic residues on their IBS similar to or higher than those in AtxA. Besides providing the basis for electrostatic interactions, the presence of anionic phospholipids in PC vesicles may facilitate non-polar interactions as a result of membrane perturbation (Buckland & Wilton, 2000).

Despite the remarkable impact of Trp at position 31 on interfacial binding of Atxs and other sPLA$_2$s, such as hGIIA (Beers et al., 2003), our subsequent site-directed mutagenesis studies revealed that the positive influence of Trp on membrane binding of sPLA$_2$s may be significantly diminished, depending on the delicate balance of contributions of each IBS residue to interfacial binding. While Atxs and hGIIA sPLA$_2$ do not contain a Trp on their IBS, DPLA$_2$, a weakly neurotoxic sPLA$_2$ from Russell's viper that differs from AtxA in only 22 residues, has a Trp residue at position 31 and yet displays interfacial binding and kinetic properties similar to those of AtxA (Petan et al., 2005; Prijatelj et al., 2003, 2008). Besides V31W, DPLA$_2$ contains several other substitutions of equivalent AtxA IBS residues: L19I, T20P, F24S, S67N, T70S, R118M and N119L (Petan et al., 2005; Prijatelj et al., 2008). By analysing the properties of a range of AtxA/DPLA$_2$ mutants and chimeras (Prijatelj et al., 2008), we were able to pinpoint the IBS residues that are crucial for the intriguing differences in interfacial binding and activity of AtxA and DPLA$_2$. Not surprisingly, the introduction of Trp31 to the AtxA/DPLA$_2$^{YIRN} chimera (see Figure 1 for sequence data) caused a very significant 50-fold increase in the rate of hydrolysis of PC vesicles. However, the impact of the same Trp residue, when introduced along with two additional substitutions, producing the AtxA^{KEW}/DPLA$_2$^{YIRN} mutant, was almost completely abolished, resulting in a low level of activity on PC vesicles, only 2 to 4-fold higher than those of AtxA/DPLA$_2$^{YIRN}, AtxA and DPLA$_2$. These results clearly confirmed our earlier suggestion (Petan et al., 2005) that the well-known positive impact of Trp31 on the interfacial binding and enzymatic activity of mammalian and venom sPLA$_2$s is, to a large extent, counterbalanced by the presence of Lys7 and Glu11 on the edges of the IBS in the case of DPLA$_2$, instead of the hydrophobic Met7 and nonpolar Gly11 in AtxA. This is most probably a consequence of altered orientation of DPLA$_2$ on the membrane, due to strong electrostatic interactions between Lys7 and Glu11 and the zwitterionic membrane surface, which prevents the productive interaction of Trp31, and thus $DPLA_2$, with the interface. Similarly, two basic residues at nearly equivalent positions in the hGIIA enzyme, Arg7 and Lys10, were shown to have a negative influence on interfacial binding (Bezzine et al., 2002; Snitko et al., 1997). In agreement with this, we have shown that the substitution of Arg at position 72 of the IBS of Atxs and $DPLA_2$ has a positive impact on interfacial binding and activity only upon introduction of a hydrophobic (Ile), but not a polar (Ser), acidic (Glu) or other basic (Lys) residue (Ivanovski et al., 2004). An additional, albeit smaller, negative effect on interfacial binding and enzymatic activity of $DPLA_2$ is provided by the presence of the polar residue Ser24 instead of the aromatic and hydrophobic Phe in AtxA (Petan et al., 2002, 2005). However, there are several residues on the IBS of $DPLA_2$, but not Atxs, that significantly improve its interfacial binding and activity. The YIRN cluster in the C-terminal part of AtxA, shown to be important for its neurotoxic effect (Ivanovski et al., 2000), contains the hydrophilic IBS residues Arg118 and Asn119, while the hydrophobic Tyr115 and Ile116 are not part of the presumed IBS. The introduction of the YIRN cluster into $DPLA_2$, instead of its KKML cluster, or into the various $AtxA/DPLA_2$ variants, resulted in a substantial decrease of the initial rates of hydrolysis of charge-neutral PC vesicles (Table 2). Therefore, the presence of a hydrophobic or aromatic IBS residue at positions 118 and 119 has a significant positive influence on interfacial binding to anionic and, particularly, zwitterionic phospholipid surfaces. Accordingly, this positive impact is evident in the case of Met118 and Leu119 in $DPLA_2$ and the $AtxA^{KKML}$ mutant, as well as of Met118 and Tyr119 in AtxB, in contrast to the hydrophilic Arg118 and Asn119 in AtxA and mutants containing the YIRN cluster (Ivanovski et al., 2000; Petan et al., 2005; Prijatelj et al., 2003, 2008). Considering the similar activities on PC-rich vesicles displayed by AtxB and the $AtxA^{KKML}$ mutant, it is obvious that the roles of Tyr and Leu at position 119 are in fact very similar. This indicates that the removal of the polar cluster Arg118/Asn119 from AtxA has a greater positive effect on interfacial binding than the aromatic or hydrophobic nature of the substituting residue at position 119, which is in accordance with the negligible impact of the substitution Y119W on interfacial binding of hGIIA $sPLA_2$ to 1,2-dioleoyl-sn-glycero-3-phosphocholine (DOPC) vesicles (Beers et al., 2003). Thus, the role of aromatic residues, including Trp, in interfacial binding of $sPLA_2s$, depends strongly, as well as on their nature, position on the IBS, and side-chain orientation, on the counterbalancing effects of polar/electrostatic interactions provided by hydrophilic/charged residues on or near the IBS. In conclusion, the most significant differences in interfacial binding and enzymatic activity of AtxA and $DPLA_2$ are therefore a consequence of the natural substitutions that have led to significant changes in the hydrophobic/aromatic character of residues on or near the IBS – M7K, G11E, F24S, V31W, R118M and N119L.

5. Interaction of Atxs with high-affinity binding proteins

Several high-affinity binding proteins for Atxs have been identified in porcine cerebral cortex. Two of these are membrane proteins of 25 kDa (R25) and 180 kDa (R180). R25, which is located in the mitochondrial membrane (Šribar et al., 2003a), binds only Atxs (Vučemilo et al., 1998), whereas R180, identified as the plasma membrane M-type $sPLA_2$ receptor ($sPLA_2R$), binds both toxic and non-toxic $sPLA_2s$ of groups IB and IIA (Čopič et al., 1999; Vardjan et al., 2001). Surprisingly, several cytosolic, high-affinity binding proteins for Atxs were identified as well – calmodulin (CaM) (Šribar et al., 2001), the γ and

sPLA$_2$	Specific activity (μmol/(min × mg))				
	POPG	POPS	POPC	10% POPS/POPC	30% POPS/POPC
AtxA	1042 ± 160	1251 ± 188	3.8 ± 0.5	56 ± 6	450 ± 21
AtxB	1149 ± 34	1189 ± 153	14 ± 2	240 ± 22	1133 ± 150
AtxC	1116 ± 91	477 ± 101	1.9 ± 0.2	14.0 ± 1.3	166 ± 21
AtnI$_2$	1070 ± 150	57 ± 5	12.3 ± 1.7	49 ± 2	159 ± 16
AtxA^{KKML}	1148 ± 126	1252 ± 40	19 ± 1	400 ± 22	1322 ± 42
AtxA^{V31W}	2102 ± 88	1964 ± 154	102 ± 7	525 ± 28	1957 ± 36
AtxA^{F24W}	914 ± 43	906 ± 56	4.4 ± 0.5	50 ± 1	209 ± 6
AtxA^{F24Y}	822 ± 41	1304 ± 102	2.1 ± 0.1	9.5 ± 0.5	163 ± 11
AtxA^{F24N}	780 ± 29	535 ± 43	1.09 ± 0.03	5.1 ± 0.5	68 ± 3
AtxA^{F24A}	813 ± 107	1291 ± 65	1.2 ± 0.1	4.4 ± 0.3	92 ± 5
AtxA^{F24S}	400 ± 73	430 ± 42	0.71 ± 0.06	4.4 ± 0.1	24 ± 4
AtxA^{R72E}	438 ± 26	n. d.	0.33 ± 0.07	n. d.	n. d.
AtxA^{R72I}	1655 ± 108	n. d.	6.9 ± 0.8	n. d.	n. d.
AtxA^{R72K}	686 ± 102	n. d.	1.2 ± 0.1	n. d.	n. d.
AtxA^{R72S}	800 ± 82	n. d.	1.6 ± 0.3	n. d.	n. d.
DPLA$_2$	1600 ± 130	1195 ± 47	1.1 ± 0.1	88 ± 9	442 ± 31
DPLA$_2$^{YIRN}	1100 ± 120	996 ± 129	0.07 ± 0.02	4.2 ± 0.4	141 ± 20
AtxA/DPLA$_2$	1300 ± 140	n. d.	3.14 ± 0.05	n. d.	n. d.
AtxA^{KEW}/DPLA$_2$^{YIRN}	1200 ± 170	n. d.	2.2 ± 0.5	n. d.	n. d.
AtxA^{KE}/DPLA$_2$^{YIRN}	830 ± 80	n. d.	0.14 ± 0.01	n. d.	n. d.
AtxA^{W}/DPLA$_2$^{YIRN}	1900 ± 210	n. d.	26 ± 3	n. d.	n. d.
AtxA/DPLA$_2$^{YIRN}	1000 ± 120	n. d.	0.53 ± 0.06	n. d.	n. d.
AtnL	No activity	No activity	No activity	n. d.	n. d.
AtnL^{YVGD}	225 ± 35	91 ± 7	~0.005	0.49 ± 0.04	43 ± 2
AtnL^{YWGD}	180 ± 25	89 ± 10	0.22 ± 0.03	8.3 ± 0.3	96 ± 5
12-AtxA	5.2 ± 0.7	n. d.	0.76 ± 0.06	n. d.	n. d.
I-AtxA	4.0 ± 0.2	n. d.	0.96 ± 0.04	n. d.	n. d.
P-AtxA	1.29 ± 0.04	n. d.	0.13 ± 0.03	n. d.	n. d.
hGIIA	220 ± 90	40 ±18	Lag, 0.7 ± 0.2	n. d.	n. d.
hGX	14 ± 0.8	4 ± 2	30 ± 0.2	n. d.	n. d.

Table 2. Specific enzymatic activity of sPLA$_2$s on phospholipid vesicles. The specific enzymatic activities were determined on extruded phospholipid vesicles composed of 1-palmitoyl-2-oleoyl-*sn*-glycero-3-phosphocholine (POPC), 1-palmitoyl-2-oleoyl-*sn*-glycero-3-phosphoserine (POPS), 1-palmitoyl-2-oleoyl-*sn*-glycero-3-phosphoglycerol (POPG), and on vesicles with mixed compositions as stated (in molar percentage) in the Table. The initial rates of phospholipid hydrolysis were measured using a sensitive fluorescence, fatty acid displacement assay; the enzymatic activities were calculated after calibrating the responses with known amounts of oleic acid (Petan et al., 2005). The enzymatic activity value for each sPLA$_2$ is the mean ± S.D. of at least five independent measurements. Data taken from Ivanovski et al. (2004), Petan et al. (2005, 2007), Prijatelj et al. (2006b, 2008), and Singer et al. (2002).

ε isoforms of 14-3-3 proteins (Šribar et al., 2003b), and protein disulphide isomerase (Šribar et al., 2005). This suggests the possibility of not only intracellular, but also cytosolic, action of Atxs and other sPLA$_2$s. They have since been shown to be surprisingly stable in the reducing environment of the eukaryotic cytosol. They are enzymatically active at low micromolar concentrations of Ca^{2+} ions, and they are also structurally stabilized and activated by binding to CaM (Kovačič et al., 2009, 2010; Petan et al., 2005; Petrovič et al.,

2005). Atxs were also found to bind factor Xa (FXa) and display anticoagulant activity (Prijatelj et al., 2006a). Given that the topic of neuronal high-affinity binding proteins for Atxs and their putative roles in ß-neurotoxicity has been reviewed recently (Pungerčar & Križaj, 2007), here we only summarize the results obtained with a range of Atx mutants for binding to CaM, R25, FXa and the M-type sPLA2 receptor.

5.1 The C-terminal region of Atxs has a novel calmodulin binding motif

Our protein engineering and competition binding studies using a range of Atxs, DPLA2 and AtnI2 mutants and chimeras, as well as synthetic peptides corresponding to different regions of Atxs, ultimately resulted in the identification of a novel binding motif for the ubiquitous, highly conserved Ca^{2+}-sensor CaM (Kovačič et al., 2010; Prijatelj et al., 2003). The C-terminal region of AtxA, with a distinctive hydrophobic patch within the region 107–125 surrounded by several basic residues, was identified as the most important element for binding to the N-terminal methionine-rich pocket of CaM in its Ca^{2+}-induced dumbbell conformation. This work was based on the intriguing 150-fold lower lethal potency and the correspondingly 50-fold lower CaM-binding affinity of DPLA2 compared to AtxA (Prijatelj et al., 2003). Most notably, the DPLA2[YIRN] mutant, containing the C-terminal YIRN cluster of AtxA (see above and Figure 1), displayed a 7-fold increase in binding affinity for CaM, only 7-fold lower than that of AtxA. This result suggested a major role of this region for the interaction with CaM. On the other hand, introduction of the whole N-terminal half of AtxA to DPLA2 and DPLA2[YIRN] caused only a slight, 2 to 3-fold increase in binding affinities for CaM, suggesting that the N-terminal part of AtxA has only a supporting role in the binding and is probably located on the periphery of the CaM binding site (Prijatelj et al., 2008). Very recently, we confirmed unambiguously the crucial importance of the C-terminal region of Atxs in CaM binding by building several three-dimensional structural models of sPLA2–CaM complexes, based on AtxA–CaM interaction site mapping, our previous mutagenesis data and protein-docking algorithms (Kovačič et al., 2010). According to the model, the closed conformation of CaM forms a "clamp" around AtxA, with most of the C-terminal residues of AtxA in the region 108-131 being in direct contact with CaM (Figure 2, C). There are only a few additional, but important, contacts between the proteins, formed mainly by basic residues in α-helices C (Arg43 and Asn54) and D (Arg94, Lys108, Asn109 and Lys111) of AtxA. Importantly, the model was validated by superimposing the structures of several mammalian sPLA2s on that of AtxA in the complex and correctly predicting the favourable interactions of group V and X mammalian sPLA2s and CaM. The model also showed the basis for the unfavourable interactions, which prevented complex formation between the structurally very similar mammalian group IB and IIA sPLA2s and CaM. Formation of the complexes was confirmed by competition binding experiments and the augmentation of the enzymatic activity of those sPLA2s interacting with CaM. Most importantly, we have clearly demonstrated that CaM stabilizes and protects Atxs from denaturation in the reducing environment of the cell cytosol and acts as a non-essential activator of Atxs and of the mammalian group V and X sPLA2s. The fact that CaM-binding results in stabilization and augmentation of enzymatic activity of sPLA2s, ranging from the neurotoxic Atxs to the mammalian group V and X sPLA2s, suggests a novel role for CaM as a regulator of the intracellular actions of some toxic and non-toxic sPLA2s, which may be active even in the reducing environment of the cytosol (Kovačič et al., 2009, 2010). The significance of CaM-sPLA2 interactions in the pharmacological and (patho)physiological roles of sPLA2s remain to be confirmed.

5.2 The N- and C-terminal regions of Atxs are involved in binding to the neuronal M-type sPLA$_2$ receptor (R180)

The M-type sPLA$_2$ receptor, which was cloned and revealed as a member of the C-type lectin superfamily, has been well-characterized and is currently the best-known sPLA$_2$ binding target. It was shown recently that the M-type receptor is involved in regulating cell senescence (Augert et al., 2009) and is the main antigen target in autoimmune human membranous nephropathy (Beck et al., 2009). However, the pathophysiological implications of the interaction between the M-type sPLA$_2$ receptors and both toxic and non-toxic sPLA$_2$s in general are intriguing and still unknown (Lambeau & Gelb, 2008; Pungerčar & Križaj, 2007; Rouault et al., 2006). Although it has been suggested that the neuronal M-type sPLA$_2$ receptor located on the plasma membrane could be responsible for specific targeting and internalization of sPLA$_2$ neurotoxins in presynaptic nerve terminals, the results of our mutagenesis studies have shown that it may not be involved in the neurotoxicity of Atxs (Prijatelj et al., 2006b). In order to prevent the interaction between AtxA and the receptor, we prepared three mutants of AtxA, containing either the 12-amino acid-long peptide ARIRARGSIEGR, named 12-AtxA, or one each of two variants of a shorter 5-amino acid peptide (ASIGQ and ASPGQ; named I-AtxA and P-AtxA, respectively) fused to the N-terminus of AtxA. These peptides are similar to the N-terminal propeptides present in proenzymes of mammalian group IB (e.g., DSGISPR in human pancreatic proPLA$_2$) and group X sPLA$_2$s (e.g., EASRILRVHRR in the human proenzyme), which do not bind to the M-type sPLA$_2$ receptor (Hanasaki & Arita, 2002; Lambeau et al., 1995). The presence of the fusion peptides in AtxA indeed completely abolished the interaction of the toxin with the M-type receptor. However, the mutants displayed only one order of magnitude lower lethality in mice and were able to induce neurotoxic effects on a mouse phrenic nerve-hemidiaphragm preparation (see Table 1). Although the N-terminal fusion peptides acted effectively as "propeptides" for AtxA and lowered specific enzymatic activity to ~1% of the wild-type enzyme, their similar neurotoxic profiles on the neuromuscular junction indicate that minimal enzymatic activity suffices for presynaptic toxicity of sPLA$_2$s. Additionally, antibodies targeting the sPLA$_2$-binding C-type lectin-like domain 5 of the M-type sPLA$_2$ receptor were unable to abolish the neurotoxic action of AtxA on the neuromuscular preparation. Thus, the interaction of AtxA with the neuronal M-type sPLA$_2$ receptor, R180, is apparently not essential for its presynaptic neurotoxicity. Since we performed the binding experiments with the N-terminal fusion AtxA mutants, using porcine sPLA$_2$ receptor, the question still remains as to whether the fusion peptides would also abolish the binding of AtxA to the mouse R180 receptor.

Nevertheless, our site-directed mutagenesis studies have provided important information about the parts of the AtxA molecule involved in binding to R180. Interestingly, the C-terminal YIRN cluster, which is essential for neurotoxicity and binding to CaM and R25, appears to be involved in binding to R180 as well. The first clues were provided by the lower affinity of the AtxA^{KKML} mutant and DPLA$_2$ for binding to R180 (Ivanovski et al., 2000). (Both contain KKML instead of the YIRN cluster.) The higher binding affinity of mutants containing the YIRN cluster (DPLA$_2$^{YIRN} or AtxA/DPLA$_2$^{YIRN}) confirmed that it is an important part of the AtxA/R180 interaction surface. However, other parts of the molecule also contribute considerably to the binding affinity of AtxA. Interestingly, the N-terminal part of AtxA in the chimera AtxA/DPLA$_2$ did not increase the binding affinity of DPLA$_2$ to R180. However, the introduction of three N-terminal DPLA$_2$ residues (K, E and/or W) to AtxA/DPLA$_2$^{YIRN} caused an increase in the binding affinity for R180 up to the level of AtxA, confirming the expected involvement of the N-terminal part of the molecule in binding to R180 as well. Therefore, the interaction surface of AtxA with its neuronal M-type receptor, R180, extends from the C-terminal top of the molecule, across the area close to the Ca^{2+}-binding loop and the IBS, reaching the N-terminal helix on the edges of the IBS.

5.3 Atxs bind to the mitochondrial receptor R25 through their C-terminal region

The C-terminal region of Atxs appears to be essential also for binding to the, as yet unidentified, mitochondrial membrane receptor R25. The affinities for binding of the chimeric mutants to R25 ranged from that of the non-toxic AtnI$_2$, which does not bind to R25, to those of AtxA, AtnI$_2$/AtxA^{K108N} and AtnI$_2$^{N24F}/AtxA^{K108N}, which were similar to that of AtxA (Prijatelj et al., 2002). Additionally, substitutions of Phe24 in the N-terminal part of AtxA had a negligible effect on its affinity for R25, but nevertheless indicated that this residue is in the vicinity of the toxin-receptor binding surface, most probably located in the C-terminal part of AtxA (Petan et al., 2002). Furthermore, while DPLA$_2$ and the chimera AtxA/DPLA$_2$ could not completely inhibit the binding of AtxC to R25 in competition binding experiments, the substitution of their KKML cluster with the YIRN cluster led to complete inhibition with a relatively high binding affinity. This implies that residues in the region 115-119 are very important for effective binding to R25, which is apparently present in multiple isoforms. However, the complete inhibition of binding of radiolabelled AtxC to R25 by the AtxA^{KKML} mutant, in contrast to DPLA$_2$ and AtxA/DPLA$_2$, indicated that the interaction surface extends beyond the C-terminal region. Indeed there was a slight increase in binding affinity upon introduction of the N-terminal Lys7, Glu11 and/or Trp31 of DPLA$_2$ to the AtxA/DPLA$_2$^{YIRN} chimera (K, E and/or W; Table 1). While our site-directed mutagenesis and competition binding studies have so far revealed a central role for the C-terminal residues of Atxs in high affinity binding to R25, the interaction surface probably extends beyond the C-terminal region.

5.4 Atxs exert anticoagulant effects by binding to factor Xa

In the analysis of non-neurotoxic effects of Atxs, our study of an array of site-directed mutants (Prijatelj et al., 2006a) showed that basic residues in the C-terminal and ß-structure regions are important for AtxA binding to human blood coagulation factor Xa (Figure 2, D). Hence, AtxA has anticoagulant activity. In addition, the 10-fold lower affinity of AtxC for factor Xa, in comparison with AtxA, is due to the (natural) charge-reversal substitution of Lys128 by Glu, leading to a small local conformational change in the C-terminal end of AtxC, which perturbs the interaction with factor Xa (Saul et al., 2010).

6. Restoration of enzymatic activity reduces the Ca^{2+}-independent membrane damage induced by the myotoxic ammodytin L

Ammodytin L (AtnL) is a myotoxic and enzymatically inactive structural homologue of Atxs (Petan et al., 2007; Pungerčar et al., 1990). The main structural feature of snake venom sPLA$_2$-like myotoxins is the substitution of the highly-conserved aspartic acid residue at position 49, effectively preventing Ca^{2+} binding and thus abolishing catalytic activity (Petan et al., 2007; Ward et al., 2002). The most common substitution is Lys, but AtnL is one of the two known Ser49 sPLA$_2$ homologues. In addition to this replacement, several other substitutions of highly-conserved residues in the enzymatically inactive sPLA$_2$s are found concentrated in the region of the Ca^{2+}-binding loop, where the carbonyl oxygen atoms of Tyr28, Gly30 and Gly32 provide three additional coordinative bonds for Ca^{2+}-binding. The myotoxic activity of the Lys49 sPLA$_2$s, as well as of the Ser49-containing AtnL, is best characterized by their ability to induce myonecrosis *in vivo* and show a potent Ca^{2+}-independent membrane-damaging activity *in vitro* (Lomonte et al., 2009). Although the involvement of a skeletal muscle protein acceptor in the process of sPLA$_2$-induced myotoxicity cannot be ruled out, a considerable body of evidence has been accumulated indicating that the myotoxic effects are a consequence of a Ca^{2+}-independent protein-lipid interaction that causes direct damage of the plasma membrane of target muscle cells and may allow entry of extracellular Ca^{2+} and irreversible cell damage (Cintra-Francischinelli et al., 2010; Lomonte et al., 2009; Rufini et al., 1992). The calcium-independent, and thus enzymatic activity-independent, mechanism of membrane damage is also supported by experiments in which inhibition of certain enzymatically active Asp49 myotoxic sPLA$_2$s did not eliminate their myotoxic activity (Soares et al., 2001).

We recently described the first example of restoration of activity in AtnL or in any enzymatically inactive Ser49/Lys49 sPLA$_2$ homologue (Petan et al., 2007). We prepared two enzymatically active quaternary mutants of AtnL (H28Y/L31V,W/N33G/S49D), differing at position 31. Although Asn33 is not directly involved in Ca^{2+} coordination, we suspected that it might have a negative impact on the optimal local structure of the Ca^{2+}-binding loop. We therefore replaced it with a glycine residue, which is very often found at this position in catalytically active sPLA$_2$s, as well as in Atxs. Val31 was present in the AtnL^{YVGD} mutant in order to recreate the Ca^{2+}-binding loop of Atxs, while the AtnL^{YWGD} mutant had Trp31 in order to enhance its membrane binding affinity, and thus enzymatic activity, as previously shown in the case of the AtxA^{V31W} mutant described above. The successful restoration of enzymatic activity by such a small number of substitutions indicates that, apart from the residues involved in Ca^{2+} coordination, the remainder of the substrate-binding and catalytic network of AtnL is very well conserved. Apparently, in evolution, Lys49 and Ser49 sPLA$_2$ myotoxins have lost their Ca^{2+}-binding ability and enzymatic activity through subtle changes in the Ca^{2+}-binding network alone, without affecting the rest of the catalytic machinery. Our results strongly suggest that these changes were selected for their Ca^{2+}-independent membrane-damaging ability and increase the specificity of their myotoxic activity. Although the restoration of enzymatic activity in AtnL increased its cytotoxic potency and lethality in general, it had a negative effect on its Ca^{2+}-independent mechanism of membrane damage and on its ability to specifically target differentiated muscle cells *in vitro*. Given that AtnL shares a high level of identity with AtxA (74%), it was not surprising that the enzymatically active mutants of AtnL displayed a combination of the properties of both toxins. In other words, the restoration of enzymatic activity of AtnL reduced its ability to act as a potent and specific myotoxic molecule. The latter supports the idea of an evolutionary specialization of Ser49/Lys49 sPLA$_2$ homologues to perform the role of abundant and weakly lethal, but specific, muscle-targeting toxins in the arsenal of pharmacologically active molecules found in snake venom.

7. Future perspectives

Recently, we have been able to introduce a free cysteine residue into the molecule of AtxA by substituting Asn79 by Cys at the turn of the antiparallel ß-sheet. This mutation allowed for specific labelling of the neurotoxin molecule by either fluorescent or nanogold probes (Pražnikar et al., 2008, 2009). The conjugates were less toxic than the wild-type toxin, but retained its basic properties. They have proved to be a valuable tool in the study of presynaptic neurotoxicity. As an illustration, we have provided evidence that AtxA, a neurotoxic snake venom sPLA2, is indeed internalized into mammalian (mouse) motor nerve terminals (Logonder et al., 2009).

We believe that similar approaches can be used in further protein engineering of Atxs and also other ß-ntxs, especially in monitoring their trafficking and cellular action (enzymatic activity, binding to protein target molecules). This should finally lead to a complete understanding of the molecular mechanism of presynaptic toxicity of sPLA2 neurotoxins, which may also open the way for better medical treatment of snake envenomation. Moreover, similar research approaches can also be exploited in the studies of homologous mammalian sPLA2s whose biological roles are still to be clarified.

8. Acknowledgment

The authors wish to thank Dr Roger H. Pain for critical reading of the manuscript.

9. References

Augert, A., Payré, C., De Launoit, Y., Gil, J., Lambeau, G. & Bernard, D. (2009). The M-type receptor PLA2R regulates senescence through the p53 pathway. *EMBO Reports*, Vol. 10, No. 3, pp. 271-277

Beck, L.H., Bonegio, R.G.B., Lambeau, G., Beck, D.M., Powell, D.W., Cummins, T.D., Klein, J.B. & Salant, D.J. (2009). M-type phospholipase A2 receptor as target antigen in idiopathic membranous nephropathy. *New England Journal of Medicine*, Vol. 361, No. 1, pp. 11-21

Beers, S.A., Buckland, A.G., Giles, N., Gelb, M.H. & Wilton, D.C. (2003). Effect of tryptophan insertions on the properties of the human group IIA phospholipase A2: Mutagenesis produces an enzyme with characteristics similar to those of the human group V phospholipase A2. *Biochemistry*, Vol. 42, No. 24, pp. 7326-7338

Berg, O.G., Gelb, M.H., Tsai, M.D. & Jain, M.K. (2001). Interfacial enzymology: The secreted phospholipase A2-paradigm. *Chemical Reviews*, Vol. 101, No. 9, pp. 2613-2654

Bezzine, S., Koduri, R.S., Valentin, E., Murakami, M., Kudo, I., Ghomashchi, F., Sadilek, M., Lambeau, G. & Gelb, M.H. (2000). Exogenously added human group X secreted phospholipase A2 but not the group IB, IIA, and V enzymes efficiently release arachidonic acid from adherent mammalian cells. *Journal of Biological Chemistry*, Vol. 275, No. 5, pp. 3179-3191

Bezzine, S., Bollinger, J.G., Singer, A.G., Veatch, S.L., Keller, S.L. & Gelb, M.H. (2002). On the binding preference of human groups IIA and X phospholipases A2 for membranes with anionic phospholipids. *Journal of Biological Chemistry*, Vol. 277, No. 50, pp. 48523-48534

Birts, C.N., Barton, C.H. & Wilton, D.C. (2010). Catalytic and non-catalytic functions of human IIA phospholipase A$_2$. *Trends in Biochemical Sciences*, Vol. 35, No. 1, pp. 28–35

Buckland, A.G. & Wilton, D.C. (2000). Anionic phospholipids, interfacial binding and the regulation of cell functions. *Biochimica et Biophysica Acta*, Vol. 1483, No. 2, pp. 199–216

Cintra-Francischinelli, M., Pizzo, P., Angulo, Y., Gutiérrez, J.M., Montecucco, C. & Lomonte, B. (2010). The C-terminal region of a Lys49 myotoxin mediates Ca^{2+} influx in C2C12 myotubes. *Toxicon*, Vol. 55, No. 2-3, pp. 590–596

Čopič, A., Vučemilo, N., Gubenšek, F. & Križaj, I. (1999). Identification and purification of a novel receptor for secretory phospholipase A$_2$ in porcine cerebral cortex. *Journal of Biological Chemistry*, Vol. 274, No. 37, pp. 26315–26320

Gelb, M.H., Cho, W. & Wilton, D.C. (1999). Interfacial binding of secreted phospholipases A$_2$: More than electrostatics and a major role for tryptophan. *Current Opinion in Structural Biology*, Vol. 9, No. 4, pp. 428–432

Ghomashchi, F., Lin, Y., Hixon, M.S., Yu, B.Z., Annand, R., Jain, M.K. & Gelb, M.H. (1998). Interfacial recognition by bee venom phospholipase A$_2$: Insights into nonelectrostatic molecular determinants by charge reversal mutagenesis. *Biochemistry*, Vol. 37, No. 19, pp. 6697–6710

Han, S.K., Kim, K.P., Koduri, R., Bittova, L., Munoz, N.M., Leff, A.R., Wilton, D.C., Gelb, M.H. & Cho, W. (1999). Roles of Trp31 in high membrane binding and proinflammatory activity of human group V phospholipase A$_2$. *Journal of Biological Chemistry*, Vol. 274, No. 17, pp. 11881–11888

Hanasaki, K. & Arita, H. (2002). Phospholipase A$_2$ receptor: A regulator of biological functions of secretory phospholipase A$_2$. *Prostaglandins & Other Lipid Mediators*, Vol. 68-69, pp. 71–82

Ivanovski, G., Čopič, A., Križaj, I., Gubenšek, F. & Pungerčar, J. (2000). The amino acid region 115-119 of ammodytoxins plays an important role in neurotoxicity. *Biochemical and Biophysical Research Communications*, Vol. 276, No. 3, pp. 1229–1234

Ivanovski, G., Petan, T., Križaj, I., Gelb, M.H., Gubenšek, F. & Pungerčar, J. (2004). Basic amino acid residues in the ß-structure region contribute, but not critically, to presynaptic neurotoxicity of ammodytoxin A. *Biochimica et Biophysica Acta*, Vol. 1702, No. 2, pp. 217–225

Kini, R.M. & Evans, H.J. (1989). A model to explain the pharmacological effects of snake venom phospholipases A$_2$. *Toxicon*, Vol. 27, No. 6, pp. 613–635

Kini, R.M. & Chan, Y.M. (1999). Accelerated evolution and molecular surface of venom phospholipase A$_2$ enzymes. *Journal of Molecular Evolution*, Vol. 48, No. 2, pp. 125–132

Kini, R.M. (2003). Excitement ahead: Structure, function and mechanism of snake venom phospholipase A$_2$ enzymes. *Toxicon*, Vol. 42, No. 8, pp. 827–840

Kovačič, L., Novinec, M., Petan, T., Baici, A. & Križaj, I. (2009). Calmodulin is a nonessential activator of secretory phospholipase A$_2$. *Biochemistry*, Vol. 48, No. 47, pp. 11319–11328

Kovačič, L., Novinec, M., Petan, T. & Križaj, I. (2010). Structural basis of the significant calmodulin-induced increase in the enzymatic activity of secreted phospholipases A$_2$. *Protein Engineering, Design & Selection*, Vol. 23, No. 6, pp. 479–487

Križaj, I. (2011). Ammodytoxin: A window into understanding presynaptic toxicity of secreted phospholipases A$_2$ and more. *Toxicon*, Vol. 58, No. 3, pp. 219–229

Lambeau, G., Ancian, P., Nicolas, J.P., Beiboer, S.H., Moinier, D., Verheij, H. & Lazdunski, M. (1995). Structural elements of secretory phospholipases A$_2$ involved in the binding to M-type receptors. *Journal of Biological Chemistry*, Vol. 270, No. 10, pp. 5534-5540

Lambeau, G. & Gelb, M.H. (2008). Biochemistry and physiology of mammalian secreted phospholipases A2. *Annual Review of Biochemistry*, Vol. 77, pp. 495-520

Lee, B.I., Yoon, E.T. & Cho, W. (1996). Roles of surface hydrophobic residues in the interfacial catalysis of bovine pancreatic phospholipase A2. *Biochemistry*, Vol. 35, No. 13, pp. 4231-4240

Lin, Y., Nielsen, R., Murray, D., Hubbell, W.L., Mailer, C., Robinson, B.H. & Gelb, M.H. (1998). Docking phospholipase A2 on membranes using electrostatic potential-modulated spin relaxation magnetic resonance. *Science*, Vol. 279, No. 5358, pp. 1925-1929

Logonder, U., Jenko-Pražnikar, Z., Scott-Davey, T., Pungerčar, J., Križaj, I. & Harris, J.B. (2009). Ultrastructural evidence for the uptake of a neurotoxic snake venom phospholipase A2 into mammalian motor nerve terminals. *Experimental Neurology*, Vol. 219, No. 2, pp. 591-594

Lomonte, B., Angulo, Y., Sasa, M. & Gutiérrez, J.M. (2009). The phospholipase A2 homologues of snake venoms: Biological activities and their possible adaptive roles. *Protein and Peptide Letters*, Vol. 16, No. 8, pp. 860-876

Meldolesi, J. (2002). Rapidly exchanging Ca^{2+} stores: Ubiquitous partners of surface channels in neurons. *News in Physiological Sciences*, Vol. 17, pp. 144-149

Montecucco, C., Gutiérrez, J.M. & Lomonte, B. (2008). Cellular pathology induced by snake venom phospholipase A2 myotoxins and neurotoxins: Common aspects of their mechanisms of action. *Cellular and Molecular Life Sciences*, Vol. 65, No. 18, pp. 2897-2912

Mounier, C.M., Ghomashchi, F., Lindsay, M.R., James, S., Singer, A.G., Parton, R.G. & Gelb, M.H. (2004). Arachidonic acid release from mammalian cells transfected with human groups IIA and X secreted phospholipase A2 occurs predominantly during the secretory process and with the involvement of cytosolic phospholipase A2-α. *Journal of Biological Chemistry*, Vol. 279, No. 24, pp. 25024-25038

Murakami, M., Taketomi, Y., Miki, Y., Sato, H., Hirabayashi, T. & Yamamoto, K. (2011). Recent progress in phospholipase A2 research: From cells to animals to humans. *Progress in Lipid Research*, Vol. 50, No. 2, pp. 152-192

Okeley, N.M. & Gelb, M.H. (2004). A designed probe for acidic phospholipids reveals the unique enriched anionic character of the cytosolic face of the mammalian plasma membrane. *Journal of Biological Chemistry*, Vol. 279, No. 21, pp. 21833-21840

Pan, Y.H., Yu, B.-Z., Singer, A.G., Ghomashchi, F., Lambeau, G., Gelb, M.H., Jain, M.K. & Bahnson, B.J. (2002). Crystal structure of human group X secreted phospholipase A2. Electrostatically neutral interfacial surface targets zwitterionic membranes. *Journal of Biological Chemistry*, Vol. 277, No. 32, pp. 29086-29093

Paoli, M., Rigoni, M., Koster, G., Rossetto, O., Montecucco, C. & Postle, A.D. (2009). Mass spectrometry analysis of the phospholipase A2 activity of snake pre-synaptic neurotoxins in cultured neurons. *Journal of Neurochemistry*, Vol. 111, No. 3, pp. 737-744

Petan, T., Križaj, I., Gubenšek, F. & Pungerčar, J. (2002). Phenylalanine-24 in the N-terminal region of ammodytoxins is important for both enzymic activity and presynaptic toxicity. *Biochemical Journal*, Vol. 363, Pt. 2, pp. 353-358

Petan, T., Križaj, I., Gelb, M.H. & Pungerčar, J. (2005). Ammodytoxins, potent presynaptic neurotoxins, are also highly efficient phospholipase A2 enzymes. *Biochemistry*, Vol. 44, No. 37, pp. 12535-12545

Petan, T., Križaj, I. & Pungerčar, J. (2007). Restoration of enzymatic activity in a Ser-49 phospholipase A_2 homologue decreases its Ca^{2+}-independent membrane-damaging activity and increases its toxicity. *Biochemistry*, Vol. 46, No. 44, pp. 12795–12809

Petrovič, U., Šribar, J., Pariš, A., Rupnik, M., Kržan, M., Vardjan, N., Gubenšek, F., Zorec, R. & Križaj, I. (2004). Ammodytoxin, a neurotoxic secreted phospholipase A_2, can act in the cytosol of the nerve cell. *Biochemical and Biophysical Research Communications*, Vol. 324, No. 3, pp. 981–985

Petrovič, U., Šribar, J., Matis, M., Anderluh, G., Peter-Katalinić, J., Križaj, I. & Gubenšek, F. (2005). Ammodytoxin, a secretory phospholipase A_2, inhibits G2 cell-cycle arrest in the yeast *Saccharomyces cerevisiae*. *Biochemical Journal*, Vol. 391, Pt. 2, pp. 383–388

Pražnikar, Z.J., Kovačič, L., Rowan, E.G., Romih, R., Rusmini, P., Poletti, A., Križaj, I. & Pungerčar, J. (2008). A presynaptically toxic secreted phospholipase A_2 is internalized into motoneuron-like cells where it is rapidly translocated into the cytosol. *Biochimica et Biophysica Acta*, Vol. 1783, No. 6, pp. 1129–1139

Pražnikar, Z.J., Petan, T. & Pungerčar, J. (2009). A neurotoxic secretory phospholipase A_2 induces apoptosis in motoneuron-like cells. *Annals of the New York Academy of Sciences*, Vol. 1152, pp. 215–224

Prijatelj, P., Čopič, A., Križaj, I., Gubenšek, F. & Pungerčar, J. (2000). Charge reversal of ammodytoxin A, a phospholipase A_2-toxin, does not abolish its neurotoxicity. *Biochemical Journal*, Vol. 352, Pt. 2, pp. 251–255

Prijatelj, P., Križaj, I., Kralj, B., Gubenšek, F. & Pungerčar, J. (2002). The C-terminal region of ammodytoxins is important but not sufficient for neurotoxicity. *European Journal of Biochemistry*, Vol. 269, No. 23, pp. 5759–5764

Prijatelj, P., Šribar, J., Ivanovski, G., Križaj, I., Gubenšek, F. & Pungerčar, J. (2003). Identification of a novel binding site for calmodulin in ammodytoxin A, a neurotoxic group IIA phospholipase A_2. *European Journal of Biochemistry*, Vol. 270, No. 14, pp. 3018–3025

Prijatelj, P., Charnay, M., Ivanovski, G., Jenko, Z., Pungerčar, J., Križaj, I. & Faure, G. (2006a). The C-terminal and ß-wing regions of ammodytoxin A, a neurotoxic phospholipase A_2 from *Vipera ammodytes ammodytes*, are critical for binding to factor Xa and for anticoagulant effect. *Biochimie*, Vol. 88, No. 1, pp. 69–76

Prijatelj, P., Vardjan, N., Rowan, E.G., Križaj, I. & Pungerčar, J. (2006b). Binding to the high-affinity M-type receptor for secreted phospholipases A_2 is not obligatory for the presynaptic neurotoxicity of ammodytoxin A. *Biochimie*, Vol. 88, No. 10, pp. 1425–1433

Prijatelj, P., Jenko Pražnikar, Z., Petan, T., Križaj, I. & Pungerčar, J. (2008). Mapping the structural determinants of presynaptic neurotoxicity of snake venom phospholipases A_2. *Toxicon*, Vol. 51, No. 8, pp. 1520–1529

Pungerčar, J., Liang, N.S., Štrukelj, B. & Gubenšek, F. (1990). Nucleotide sequence of a cDNA encoding ammodytin L. *Nucleic Acids Research*, Vol. 18, No. 15, p. 4601

Pungerčar, J., Križaj, I., Liang, N.S. & Gubenšek, F. (1999). An aromatic, but not a basic, residue is involved in the toxicity of group-II phospholipase A_2 neurotoxins. *Biochemical Journal*, Vol. 341, Pt. 1, pp. 139–145

Pungerčar, J. & Križaj, I. (2007). Understanding the molecular mechanism underlying the presynaptic toxicity of secreted phospholipases A_2. *Toxicon*, Vol. 50, No. 7, pp. 871–892

Ramirez, F. & Jain, M.K. (1991). Phospholipase A_2 at the bilayer interface. *Proteins: Structure, Function, and Bioinformatics*, Vol. 9, No. 4, pp. 229–239

Renetseder, R., Brunie, S., Dijkstra, B.W., Drenth, J. & Sigler, P.B. (1985). A comparison of the crystal structures of phospholipase A_2 from bovine pancreas and *Crotalus atrox* venom. *Journal of Biological Chemistry*, Vol. 260, No. 21, pp. 11627–11634

Rigoni, M., Caccin, P., Gschmeissner, S., Koster, G., Postle, A.D., Rossetto, O., Schiavo, G. & Montecucco, C. (2005). Equivalent effects of snake PLA_2 neurotoxins and lysophospholipid-fatty acid mixtures. *Science*, Vol. 310, No. 5754, pp. 1678–1680

Rigoni, M., Paoli, M., Milanesi, E., Caccin, P., Rasola, A., Bernardi, P. & Montecucco, C. (2008). Snake phospholipase A_2 neurotoxins enter neurons, bind specifically to mitochondria, and open their transition pores. *Journal of Biological Chemistry*, Vol. 283, No. 49, pp. 34013–34020

Rosenberg, P. (1997) Pitfalls to avoid in the study of correlations between enzymatic activity and pharmacological properties of phospholipase A_2 enzymes, In: *Venom Phospholipase A_2 Enzymes: Structure, Function and Mechanism* (Kini, R. M., Ed.), pp. 155-183, John Wiley & Sons, Chichester, England.

Rouault, M., Rash, L.D., Escoubas, P., Boilard, E., Bollinger, J., Lomonte, B., Maurin, T., Guillaume, C., Canaan, S., Deregnaucourt, C., Schrével, J., Doglio, A., Gutiérrez, J.M., Lazdunski, M., Gelb, M.H. & Lambeau, G. (2006). Neurotoxicity and other pharmacological activities of the snake venom phospholipase A_2 OS_2: The N-terminal region is more important than enzymatic activity. *Biochemistry*, Vol. 45, No. 18, pp. 5800-5816

Rufini, S., Cesaroni, P., Desideri, A., Farias, R., Gubenšek, F., Gutiérrez, J.M., Luly, P., Massoud, R., Morero, R. & Pedersen, J.Z. (1992). Calcium ion independent membrane leakage induced by phospholipase-like myotoxins. *Biochemistry*, Vol. 31, No. 49, pp. 12424–12430

Saul, F.A., Prijatelj-Žnidaršič, P., Vulliez-le Normand, B., Villette, B., Raynal, B., Pungerčar, J., Križaj, I. & Faure, G. (2010). Comparative structural studies of two natural isoforms of ammodytoxin, phospholipases A_2 from *Vipera ammodytes ammodytes* which differ in neurotoxicity and anticoagulant activity. *Journal of Structural Biology*, Vol. 169, No. 3, pp. 360–369

Scott, D.L., White, S.P., Otwinowski, Z., Yuan, W., Gelb, M.H. & Sigler, P.B. (1990). Interfacial catalysis: The mechanism of phospholipase A_2. *Science*, Vol. 250, No. 4987, pp. 1541–1546

Scott, D.L., Mandel, A.M., Sigler, P.B. & Honig, B. (1994). The electrostatic basis for the interfacial binding of secretory phospholipases A_2. *Biophysical Journal*, Vol. 67, No. 2, pp. 493–504

Singer, A.G., Ghomashchi, F., Le Calvez, C., Bollinger, J., Bezzine, S., Rouault, M., Sadilek, M., Nguyen, E., Lazdunski, M., Lambeau, G. & Gelb, M.H. (2002). Interfacial kinetic and binding properties of the complete set of human and mouse groups I, II, V, X, and XII secreted phospholipases A_2. *Journal of Biological Chemistry*, Vol. 277, No. 50, pp. 48535-48549

Snitko, Y., Koduri, R.S., Han, S.K., Othman, R., Baker, S.F., Molini, B.J., Wilton, D.C., Gelb, M.H. & Cho, W. (1997). Mapping the interfacial binding surface of human secretory group IIa phospholipase A_2. *Biochemistry*, Vol. 36, No. 47, pp. 14325–14333

Soares, A.M., Andrião-Escarso, S.H., Bortoleto, R.K., Rodrigues-Simioni, L., Arni, R.K., Ward, R.J., Gutiérrez, J.M. & Giglio, J.R. (2001). Dissociation of enzymatic and pharmacological properties of piratoxins-I and -III, two myotoxic phospholipases A_2 from *Bothrops pirajai* snake venom. *Archives of Biochemistry and Biophysics*, Vol. 387, No. 2, pp. 188–196

Šribar, J., Čopič, A., Pariš, A., Sherman, N.E., Gubenšek, F., Fox, J.W. & Križaj, I. (2001). A high affinity acceptor for phospholipase A_2 with neurotoxic activity is a calmodulin. *Journal of Biological Chemistry*, Vol. 276, No. 16, pp. 12493–12496

Šribar, J., Čopič, A., Poljšak-Prijatelj, M., Kuret, J., Logonder, U., Gubenšek, F. & Križaj, I. (2003a). R25 is an intracellular membrane receptor for a snake venom secretory phospholipase A_2. *FEBS Letters*, Vol. 553, No. 3, pp. 309–314

Šribar, J., Sherman, N.E., Prijatelj, P., Faure, G., Gubenšek, F., Fox, J.W., Aitken, A., Pungerčar, J. & Križaj, I. (2003b). The neurotoxic phospholipase A_2 associates, through a non-phosphorylated binding motif, with 14-3-3 protein γ and ε isoforms. *Biochemical and Biophysical Research Communications*, Vol. 302, No. 4, pp. 691–696

Šribar, J., Anderluh, G., Fox, J.W. & Križaj, I. (2005). Protein disulphide isomerase binds ammodytoxin strongly: Possible implications for toxin trafficking. *Biochemical and Biophysical Research Communications*, Vol. 329, No. 2, pp. 733–737

Stahelin, R.V. & Cho, W. (2001). Differential roles of ionic, aliphatic, and aromatic residues in membrane-protein interactions: A surface plasmon resonance study on phospholipases A_2. *Biochemistry*, Vol. 40, No. 15, pp. 4672–4678

Sumandea, M., Das, S., Sumandea, C. & Cho, W. (1999). Roles of aromatic residues in high interfacial activity of *Naja naja atra* phospholipase A_2. *Biochemistry*, Vol. 38, No. 49, pp. 16290–16297

Thouin, L.G., Ritonja, A., Gubenšek, F. & Russell, F.E. (1982). Neuromuscular and lethal effects of phospholipase A from *Vipera ammodytes* venom. *Toxicon*, Vol. 20, No. 6, pp. 1051–1058

Tzeng, M.C., Yen, C.H., Hseu, M.J., Dupureur, C.M. & Tsai, M.D. (1995). Conversion of bovine pancreatic phospholipase A_2 at a single site into a competitor of neurotoxic phospholipases A_2 by site-directed mutagenesis. *Journal of Biological Chemistry*, Vol. 270, No. 5, pp. 2120–2123

Valentin, E. & Lambeau, G. (2000). What can venom phospholipases A_2 tell us about the functional diversity of mammalian secreted phospholipases A_2? *Biochimie*, Vol. 82, No. 9–10, pp. 815–831

Vardjan, N., Sherman, N.E., Pungerčar, J., Fox, J.W., Gubenšek, F. & Križaj, I. (2001). High-molecular-mass receptors for ammodytoxin in pig are tissue-specific isoforms of M-type phospholipase A_2 receptor. *Biochemical and Biophysical Research Communications*, Vol. 289, No. 1, pp. 143–149

Vučemilo, N., Čopič, A., Gubenšek, F., & Križaj, I. (1998). Identification of a new high-affinity binding protein for neurotoxic phospholipases A_2. *Biochemical and Biophysical Research Communications*, Vol. 251, No. 1, pp. 209–212

Ward, R.J., Chioato, L., De Oliveira, A.H.C., Ruller, R. & Sá, J.M. (2002). Active-site mutagenesis of a Lys49-phospholipase A_2: Biological and membrane-disrupting activities in the absence of catalysis. *Biochemical Journal*, Vol. 362, Pt. 1, pp. 89–96

Yau, W.M., Wimley, W.C., Gawrisch, K. & White, S.H. (1998). The preference of tryptophan for membrane interfaces. *Biochemistry*, Vol. 37, No. 42, pp. 14713–14718

Yu, B.Z., Berg, O.G. & Jain, M.K. (1993). The divalent cation is obligatory for the binding of ligands to the catalytic site of secreted phospholipase A_2. *Biochemistry*, Vol. 32, No. 25, pp. 6485–6492

7

Site-Directed Mutagenesis in the Research of Protein Kinases - The Case of Protein Kinase CK2

Ewa Sajnaga, Ryszard Szyszka and Konrad Kubiński
Department of Molecular Biology, Institute of Biotechnology,
The John Paul II Catholic University of Lublin,
Poland

1. Introduction

Protein kinases constitute one of the largest and best-explored superfamilies of mammalian genes. The human genome encodes approximately 518 protein kinase genes, and the majority of their proteins have been characterized to some extent. Many of protein kinases have been implicated in signal transduction pathways that regulate growth and survival of cells, indicating their potential role in cancer (Manning et al., 2002a). Indeed, most cancers are associated with disregulation of protein kinases or by loss or damage of cellular protein kinase inhibitors (Brognard & Hunter, 2011; Johnson, 2009; Pearson & Fabbro, 2004).

A typical protein kinase (based on 478 known enzymes) is characterized by the presence of a highly homologous kinase catalytic domain of about 300 amino acid residues. This domain serves three distinct roles: (a) binding and orientation of the phosphate donor (ATP or, rarely, a different nucleoside triphosphate) in a complex with a divalent cation (usually Mg^{2+} or Mn^{2+}); (b) binding and orientation of the protein substrate; and (c) transfer of the γ-phosphate from the NTP to the hydroxyl moiety of the acceptor residue of the protein substrate.

In response to a variety of regulatory signals, protein kinases phosphorylate specific serine, threonine, or tyrosine residues within target proteins to modify their biological activities. Phosphorylation of several proteins in the cell creates recognition and/or regulatory sites that influence many properties of the target proteins (*e.g.*, catalytic activity, localization, sensitivity to proteolytic degradation, protein-protein interaction, etc). In eukaryotes, Ser/Thr and Tyr protein kinases (Note: The single- and three-letter codes for the amino acids are given in Table 1 in the chapter by Figurski et al.) play a key role in molecular networks controlling the activity of various signaling proteins (Brognard & Hunter, 2011; Cohen, 2002). Ser/Thr and Tyr-protein kinases form the largest protein family in the human genome (Pandit et al., 2004). They constitute about 2-3% of the proteomes of other model organisms, such as *Saccharomyces cerevisiae*, *Caenorhabditis elegans* and *Drosophila melanogaster* (Manning et al., 2002 a, b; Goldberg et al., 2006; Plowman et al., 1999). Based on the conserved features of catalytic domains of eukaryotic protein kinases, Hanks and coworkers have placed the kinases into various classes, groups, and subfamilies (Hanks et al., 1988).

The first 3D structure of a protein kinase was determined for PKA by X-ray crystallography. It revealed the basic bi-lobed scaffold formed by N- and C-terminal lobes that has been observed in all the protein kinase structures solved to date. The N-terminal lobe of the kinase fold comprises of an anti-parallel β-sheet made of five β-strands ($\beta1$ – $\beta5$) and a single αC-helix. The C-terminal lobe is larger and is mainly composed of α-helices. The nucleotide- and substrate-binding pockets are located in the cleft between the two lobes. The phosphate groups of ATP are positioned for phosphotransfer by their interactions with conserved residues in the N- and C-terminal lobes. These include a glycine-rich loop characterized by the GXGXXG motif (where X represents any amino acid) between the $\beta1$ and $\beta2$ strands, a Lys residue localized by a salt bridge formed by a Lys-Glu pair (K72 and E91), and Mg^{2+} ions. The conserved Asn (N171) and Asp (D184) further coordinate the metal ions. The catalytic loop situated in the C-terminal lobe contains aspartate (D166), referred to as the catalytic base that facilitates extraction of a proton from the hydroxyl side-chains of the phospho-sites of the substrates. The activation segment (20–30 residues in length) caps the C-terminal lobe. This segment forms a part of the substrate-binding pocket and shows high structural variation in the active and inactive kinase structures.

The grouping of the protein kinases based on catalytic subunit sequence similarity results in clustering of kinases that share functional features, such as preferred sites of phosphorylation, the mode of regulation and cellular localization. The similarity in the amino acid sequence of the catalytic domains of protein kinases has proven to be a good indicator of other features held in common by the different members of the family.

The diversity of essential functions mediated by kinases is shown by the conservation of approximately 50 distinct kinase families that have been identified in yeast, invertebrates, and mammals. Protein kinases can be clustered into groups, families, and sub-families. These classifications are based on sequence similarity and biochemical function. Among the 518 human protein kinases, 478 belong to a single superfamily whose catalytic domains are related in sequence.

Protein kinases are divided into 10 main groups, which organize diversity and compare genes between distant organisms (Miranda-Saavedra & Barton, 2007). The groups are named as follows: AGC, CAMK, CK1, CMGC, STE, RGC, Other, TK, TK, and Atypical. This classification was first used for characterization of the human kinome (all the kinases encoded in the genome) (Manning et al., 2002) and is based on an earlier classification by Hanks and Hunter (1995).

1.1 The CMGC kinases

This group of Ser/Thr protein kinases was named after the initials of some members (CDK, MAPK, GSK3, and CLK). It includes key kinases involved in growth, stress-response, and the cell cycle, and kinases involved in splicing and metabolic control. The four well characterized subfamilies of this group include the following: cyclin-dependent kinases (CDK) (Liolli, 2010); mitogen-activated protein kinases (MAPK) (Biondi & Nebreda, 2003; Zhang & Dong, 2007); glycogen synthase kinases (GSK) (Biondi & Nebreda, 2003); and cell kinases 2 (CK2), better known as protein kinase CK2 or casein kinase 2 (St-Denis & Litchfield, 2009). The CMGC group also contains other members. These are the following: SR protein kinases [phosphorylating serine- and arginine-rich proteins engaged in regulation of splicing and nuclear transport (Ghosh & Adams, 2011)] and DYRK protein kinases, *i.e.* dual-specificity tyrosine-phosphorylated protein kinases, presumably involved in brain development (Becker et al., 1998).

1.2 Protein kinase CK2

Protein kinase CK2, formerly called casein kinase II, is a ubiquitous second messenger-independent protein kinase found in all eukaryotic organisms examined (Jensen et al., 2007; Niefind et al., 2009; Litchfield, 2003, Kubiński et al., 2007). This enzyme, which has been studied for over 50 years, is able to phosphorylate more than 300 substrates, on serine, threonine and tyrosine (Meggio & Pinna, 2003; Vilk et al., 2008). As the list of targets for CK2 continues to grow, it is becoming evident that CK2 has the potential to participate in the regulation of various cellular processes. Most of the CK2 substrates reported so far correspond to proteins that participate in cell signaling (Ahmed et al., 2002; Meggio & Pinna, 2003).

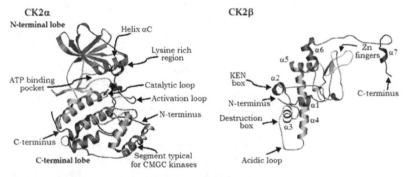

Fig. 1. Model structures of the CK2α and CK2β subunits from *Mytilus galloprovincialis* (Mediterranean mussel) (Koyanou-Koutsokou et al., 2011b)
The structural features of the CK2α and CK2β subunits were elaborated using the SWISS-MODEL Workspace for protein structure homology modeling (Arnold et al., 2006; Kopp and Schwede 2004) and 1ds5D (Batistuta et al., 2000) or 3EED (Raaf et al., 2008) as templates, respectively.

Protein kinase CK2 is distributed ubiquitously in eukaryotic organisms, where it appears as a tetrameric complex composed of two catalytic subunits (α/α') associated with a dimer of regulatory β subunits (Figs. 1 & 2). The CK2 tetramer exhibits constitutive activity that can be easily detected in most cellular or tissue extracts in the absence of any stimulatory compounds. In many organisms, distinct isoenzymic forms of the catalytic subunit of CK2 have been identified (Glover, 1998; Kolaiti et al., 2011; Kouyanou-Koutsoukou et al., 2011a, b; Maridor et al., 1991; Litchfield et al., 1990; Shi et al., 2001). In humans, only a single regulatory CK2β subunit has been identified; but multiple forms of CK2β have been identified in other organisms, such as *Saccharomyces cerevisiae* (Glover, 1998). Complementary evidence indicates that dimers of CK2β are at the core of the tetrameric CK2 complexes (Graham & Litchfield, 2000; Pinna & Meggio, 1997). Tetrameric CK2 complexes may contain identical (*i.e.*, $\alpha_2\beta_2$ or $\alpha'_2\beta_2$) or non-identical (*i.e.*, $\alpha\alpha'\beta_2$) catalytic subunits (Gietz et al., 1995). Holoenzyme composition may influence CK2 properties, namely nucleotide and protein substrate specificity and sensitivity to effectors (Janeczko et al., 2011). Protein kinase CK2 holoenzyme and its catalytic subunit alone can use both ATP and GTP as phosphate donors (Issinger, 1993).

The catalytic subunits of CK2α and CK2α′ are the products of separate genes located in different chromosomes. The 330 N-terminal amino acids exhibit over 90% sequence identity. However, the C-terminal sequences are unrelated (Olsten & Litchfield, 2004). The unique C-terminal domains of the catalytic subunits are highly conserved among species (e.g., the amino acid sequences of the C-termini of the catalytic subunits of human and chicken CK2α and CK2α′ exhibit 98% and 97% identity, respectively), indicating a possible functional importance for this domain (Litchfield, 2003).

Although there is no known difference between the catalytic activities of CK2α and CK2α′, there is evidence that they exhibit functional specialization (Duncan & Lichfield, 2008; Faust & Montenarch, 2000). The CK2α subunit is phosphorylated at C-terminal sites (Thr344, Thr360, Ser362 and Ser360) by p34[cdc2] during cell cycle progression, while CK2α′ is not phosphorylated (St-Denis et al., 2009). Further evidence to support the idea that CK2α and CK2α′ have independent functions in the cell is provided by the different specificities of cellular binding proteins, such as CKIP-1, Hsp90, Pin-1, and PP2A (Olsten et al., 2005).

Despite the many isoforms of catalytic subunits, only one regulatory subunit has been identified for CK2β in mammals (Allende and Allende, 1995). In contrast to the activity of regulatory subunits of other kinases, such as PKA (cAMP-dependent protein kinase) and CDK (cyclin-dependent protein kinase), CK2β does not switch on or off the intrinsic activity of the catalytic subunits (Bolanos-Garcia et al., 2006).

The CK2β regulatory subunit is remarkably conserved among species, but it does not have homology with the regulatory subunits of other protein kinases (Bibby & Litchfield, 2005). The amino acid sequence of the CK2β regulatory subunit is almost identical in Homo sapiens, Drosophila melanogaster, Ceratitis capitata (Mediterranean fruit fly), Danio rerio (zebrafish), Ciona intestinalis (sea squirt), and Mytilus galloprovincialis (Mediterranean mussel) (Kouyanou-Koutsoukou et al., 2011a, b; Kolaiti et al., 2011). It is completely identical in birds and mammals (Maridor et al., 1991; Wirkner et al., 1994). In contrast, the fruit fly D. melanogaster has four regulatory subunit genes. They are used for one CK2α (DmCK2α) and three CK2βs (DmCK2β, DmCK2β′ and DmCK2βtes) (Jauch et al., 2002). Zea mays has three isoforms of the catalytic α-subunit (CK2a-1, CK2a-2 and CK2a-3) and three regulatory β-subunits (CK2b-1, CK2b-2 and CK2b-3) (Riera et al., 2001). S. cerevisiae CK2 holoenzyme contains two regulatory β-subunits (β and β′). They cannot substitute for each other, and both of them are needed to form a fully active enzymatic unit (Kubinski et al., 2007).

Results presented by several groups and obtained by the use of a variety of approaches, including X-ray crystallography, have determined that a dimer of the CK2β subunits forms the core of the CK2 tetramer (Chantalat et al., 1999; Sarno et al., 2000; Canton et al., 2001).

The CK2β regulatory subunit is a compact, globular homodimer that shows high amino acid sequence conservation across species. The N-terminal domain (amino acids 1-104) is globular and contains four α-helices (marked as α1-α4 in Fig. 1). Helices α1 (residues 9–14), α2 (residues 27–31) and α3 (residues 46–54) wrap around α4 (residues 66–89) (Bolanos-Garcia et al., 2006). This part of the protein contains autophosphorylation sites, consisting of serines 2, 3, and possibly 4 (Boldyreff et al., 1993). Studies conducted by Zhang and coworkers (2002) indicate that phosphorylation of these sites enhances CK2β stability. The first 20 N-terminal amino acids of the CK2β regulatory subunit are also involved in the interaction with Nopp140, a protein that binds a nuclear localization sequence and shuttles between the nucleus and the cytoplasm (Li et al., 1997). This part of the protein also contains two motifs that have been previously characterized as motifs that regulate cyclin degradation. The CK2β regulatory subunit has a sequence resembling the nine-amino-acid motif called the destruction box, which plays a key role in the specific degradation of cyclin B at the end of mitosis (King et al., 1996). This motif, located in helix α3, contains three highly conserved residues that conform to the general destruction box consensus (RXXLXXXXN/D) (Bolanos-Garcia et al., 2006). Interestingly, this motif is located on a surface-exposed α3 helix, where it would be available for recognition by the cellular degradation machinery. A signal known as the KEN box, which was found previously in mitotic cyclins and which has been shown to play a role in mediating cell cycle-dependent protein degradation, is also present in CK2β. This degradation motif is characterized by the minimal consensus sequence KEN, but it is often followed shortly by either an N or D residue and is often preceded by another N or D residue. A similar sequence (D_{32}KFNLTGLN$_{40}$) forms helix α2 of the CK2β protein (Bibby & Litchfield, 2005).

The N-terminal part of the CK2β also contains an "acidic loop" between helices α3 and α4. This acidic, surface-exposed region of the protein, encoded by residues 55-64, has been identified as the site on CK2 that binds polyamines, which are known to stimulate CK2 activity *in vitro* (Meggio et al., 1994; Leroy et al., 1997).

The analysis of the CK2β regulatory subunit structure by X-ray crystallography revealed the importance of the zinc finger in CK2β regulatory subunit dimerization (Chantal et al., 1999). The zinc-finger region is characterized by four conserved cysteine residues (residues 109, 114, 137 and 140), which mediate the interaction that allows the CK2β dimer to form the core of the CK2 holoenzyme (Chantal et al., 1999; Canton et al., 2001).

The C-terminal part of the CK2β regulatory subunit (residues 178-205) contains a large loop (residues 178-193) and helix α7 (residues 194-200). Although helix α7 is located away from helices α1-α6, the C-terminal amino acids (190-205) contribute to the formation of the CK2β regulatory subunit dimer (Niefind et al., 2001). This part of the regulatory subunit contains two phosphorylation sites: Thr213, which is phosphorylated by the checkpoint kinase Chk1 (Kristensen et al., 2004) and Ser209, which is phosphorylated *in vitro* and in mammalian cells by p34[cdc2] in a cell-cycle-dependent manner (Litchfield et al., 1995).

The traditional view of the CK2β regulatory subunit is that it functions as a component of tetrameric CK2 complexes and that it is the regulator of the catalytic CK2α and CK2α' subunits, enhancing their stability, specificity and activity. As an example, the CK2β regulatory subunit stimulates CK2 holoenzyme activity towards certain protein substrates, such as topoisomerase II (Leroy et al., 1999), and inhibits others, like calmodulin (Marin et al., 1999).

It was shown that CK2β does not exist exclusively within stable CK2 complexes. This observation raises the prospect that CK2β has functions that are independent of its role as the regulatory subunit of CK2. For example, overexpression of CK2β in the fission yeast *Schizosaccharomyces pombe* revealed severe growth defects and a multiseptated phenotype, whereas CK2α overexpression had no effect (Roussou & Draetta, 1994).

CK2β seems to interact directly with more than 40 different proteins, including other protein kinases such as A-Raf, Chk1, Chk2, PKC-ζ, Mos and p90[rsk] (Bibby & Lichfield, 2005; Bolanos-Garcia et al., 2006; Olsen & Guerra, 2008). It was shown that association of the human protein kinases Chk1, Mos, and A-Raf is mediated by the C-terminal region of the CK2β subunit and that these associations involve some residues that interact with the catalytic CK2α subunit (Chen et al., 1997; Lieberman & Ruderman, 2004; Olsen & Guerra, 2008). The interaction between Chk1 and CK2β leads to an increase in the Cdc25C phosphorylation activity of Chk1. Screening of several cell lines has shown that the association between CK2β and Chk1 is also formed *in vivo* (Guerra at al., 2003).

Overexpression of CK2 has been linked to several pathological conditions, ranging from cardiovascular pathologies and cancer progression to neurodegenerative disorders (*e.g.*, Alzheimer's disease, Parkinson's disease, brain ischemia) and infectious diseases (Guerra & Issinger, 2008; Ahmad et al., 2008; Trembley et al., 2009). Various specific, potent small molecule inhibitors of protein kinase CK2 have been developed in recent years, including condensed polyphenolic compounds, tetrabromobenzimidazole/triazole derivatives, and indoloquinazolines (Gianoncelli et al., 2009; Pagano et al., 2008; Raaf et al., 2008). Inhibition of CK2 kinase activity by these compounds display a remarkable pro-apoptotic efficacy on a number of tumor-derived cell lines, indicating a possibility of developing novel antineoplastic drugs (Batistuta, 2009; Duncan et al., 2010; Prudent et al., 2010; Unger et al., 2004).

2. Mutagenesis in studies on protein kinase CK2

Within the last 2 decades, a number of studies have produced mutants of both CK2α and CK2β that provide a valuable, yet incomplete, basis to rationalize the biochemical features of the enzyme, i.e., its constitutive activity, dual-cosubstrate specificity, acidophilic substrate specificity and tetrameric structure (Fig. 2).

2.1 Mutagenesis of the CK2α catalytic subunit

2.1.1 Mutations of CK2α in the regions responsible for constitutive activity

A majority of protein kinases need to be activated. Phosphorylation within the kinase activation loop is the most popular mode of activation. In contrast to other known protein kinases, CK2 has constitutive activity and does not demand activation. In this case, activation is achieved by the interaction between the N-terminal tail and the activation loop in the kinase domain. The role of the N-terminal segment in stable opening of the activation loop was confirmed in mutagenesis studies (Sarno et al., 2001). In particular, the Δ2-12 CK2α mutant, in comparison with the wild-type kinase, displayed an almost complete loss of activity, which was reflected by increased Km values for ATP and the peptide substrate (from 10 to 206 μM and from 26 to 140 μM, respectively). Further experiments revealed that holoenzyme reconstitution restored the activity of the mutant to the wild-type level. This demonstrates an alternative CK2β subunit-dependent mechanism to provide constitutive activity in the case of CK2 holoenzyme (Sarno et al., 2002).

Recently, molecular dynamics (MD) simulation has been carried out in order to explore the role of the CK2α N-terminal segment in the conformational behavior of the kinase (Cristiani

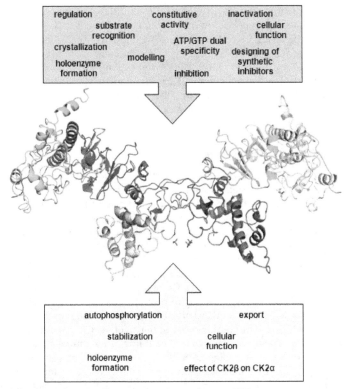

Fig. 2. Multiple applications of mutagenesis in studies on CK2.
The blue box presents various aspects of research using mutagenesis on CK2α; and the yellow box, on CK2β. The model of the human CK2 holoenzyme was developed using the PyMOL software based on the structure of the human CK2 holoenzyme (PDB code 1JWH) from the Protein Data Bank. The catalytic α subunits are presented in blue and green; the regulatory β subunits are in red and yellow.

et al., 2011). Comparison of the αC-helix RMSD (root mean square deviation) values obtained for the Δ2-12 CK2α mutant (*i.e.*, deleted for residues 2 through 12) and the wild-type kinase models show an increase in this parameter for the mutant form of the enzyme. This effect is due to instability of the CK2α conformation in the case of absence of an N-terminal segment and its interaction with the αC-helix. These results are consistent with the data presented by Sarno and collaborators, and they indicate that the complete N-terminal segment is essential for proper conformation and constitutive activity of protein kinase CK2α (Cristiani et al., 2011).

The experiment presented above is an example of the validation of *in vitro* mutagenesis studies with the use of computing analysis, but the opposite direction of studies is also possible. Two CK2α mutants, the triple mutant Y206F/R10A/Y261F and the single mutant Y125F, were constructed *in silico*. MD simulations were then carried out to study the relation between CK2 conformation and activity (Cristiani et al., 2011). The amino acids substituted in the first virtual mutant are engaged in the most important bonds between the N-terminal segment and other regions of CK2α to maintain kinase activity. The CK2α Y125F mutant is also very useful in studying the influence of Tyr125 on the conformational change of Phe121. According to Niefind and Issinger (2010), Phe121 can assume two different conformations: in and out, which regulate the activity of CK2α. Preliminary MD simulations on the two protein mutant models are very promising. The authors are currently working on the construction of both CK2 mutants. Biochemical characterization of the mutants will be carried out (Cristiani et al., 2011).

2.1.2 Mutation of CK2α in the basic regions

Protein kinase CK2 is characterized by its special aptitude to interact with negatively charged ligands. This ability correlates with the presence of several basic residues in CK2α that are not conserved in a majority of other protein kinases. These residues are located mainly in the "Lys-rich segment" and in the "p+1 loop." The Lys-rich segment ($K_{74}KKKIKR_{80}$) at the beginning of the αC-helix is a distinctive feature of CK2α (Tuazon & Traugh, 1991; Guerra et al., 1999). Results from mutational studies support the notion that this cluster is involved in substrate recognition, inhibition by heparin, down-regulation by the CK2β subunit and interaction with heat shock protein 90, and nuclear targeting (Guerra et al., 1999; Pinna & Meggio, 1997) (Table 1). CK2α mutants from *Caenorhabditis elegans* and *Xenopus laevis* (K74E/K75E and K75E/K76E, respectively) had lysines replaced by glutamic acid residues, which greatly affected the charge of this region in both mutant enzymes. The changes produced neither a significant increase in the K_m of the CK2α subunit for the casein and model peptide substrates nor changes in the affinity of the mutated CK2α subunit for the CK2β subunit during assembling a fully competent CK2 holoenzyme. The same mutations, however, had a significant effect on the affinity of CK2α for heparin and for other polyanionic inhibitors (Hu & Rubin, 1990; Gatica at al., 1994). Complete suppression of heparin inhibition was observed with the quadruple mutated K74-77A CK2α used by Vaglio and collaborators (1996). These authors showed (1) that all the four basic residues at positions 74, 75, 76, and 77 are implicated in heparin binding and (2) that the mutation of all of them was necessary to minimize heparin inhibition. Further mutagenesis studies showed that the additional basic residues cooperated with high heparin binding (apart from the 74-77 quartet). These were mainly Arg191, Arg195 and Lys198 located in the p+1 loop. However, the triple mutant for the three non-Lys-rich segment residues was less effective in heparin inhibition than was the mutant resulting from quadruple mutation of the 74-77 cluster (Vaglio et al., 1996). The triple mutant in which Lys79, Arg80 and Arg83 were changed into alanines did not alter the IC_{50} (concentration needed to give 50% inhibition) value for heparin. However, the mutant did show a reduction in the phosphorylation efficiency of the peptide substrate (and derivatives in which individual aspartyl residues were replaced by alanines). Because of these properties, it was specified that the basic residues in positions 77-83 are mainly involved in substrate recognition, rather than in heparin inhibition (Sarno et al., 1995; Vaglio et al., 1996). These authors concluded that the highly conserved 74-80 basic stretch is composed of two functionally distinct entities: (1) an N-terminal moiety mostly involved in heparin inhibition as well as in down-regulation by the β subunit and (2) the C-terminal part implicated in recognition of the crucial specificity determinant at positions n+3, but irrelevant to heparin.

Extended mutagenesis analysis combined with biochemical characterization provided clear evidence that residues responsible for both substrate recognition and down-regulation of CK2α catalytic activity are located mainly in the Lys-rich loop and p+1 loop spanning sequences 74-83 and 191-198, respectively. This corroborates the concept that the CK2β subunit down-regulates the CK2β by acting as a pseudosubstrate (Meggio et al., 1994; Sarno et al., 1996, 1997a, 1999).

Sarno and collaborators (1997b) analyzed the relative contribution of basic residues, presumably implicated in CK2-substrate interaction, in the recognition of peptide substrates varying in the number and position of acidic determinants. Sixteen derivatives of the optimal peptide substrate RRRA-DDSDDDDD, wild-type CK2 and twelve CK2α mutants defective in substrate recognition were used in the experiments. In the CK2α mutants, different basic residues implicated in substrate recognition were replaced by alanine (e.g., K49A, K74-77A, or K79A/R80A/K83A). The results obtained support the idea that the acidic residues at positions n+1 and n+3 are essential, while additional acidic residues are required for efficient phosphorylation of CK2 substrates. Kinetic analysis with CK2α mutants revealed that Lys48 was implicated in the recognition of the determinant at position n+2. Lys77 interacts with the determinants at n+3 and n+4, while Lys198 recognized the determinant at n+1 (Sarno et al., 1997b). Molecular modeling based on crystallographic data supported these observations. It showed that several of these basic residues are clustered around the active site, where they make contact with individual acidic residues of the peptide substrate, polyanionic inhibitors, regulatory elements present in the β subunit, N-terminal segment of the CK2α and possibly other proteins interacting with CK2 (Sarno et al., 1999).

2.1.3 Mutations of CK2α in catalytic subdomains

Subdomains II and VII of CK2α involved in nucleotide binding and phosphotransfer are in close proximity to each other in the three-dimensional structure. CK2α differs from more than 95% of other known protein kinases in having Val66 instead of the corresponding alanine within conserved region II and Trp176 instead of the corresponding phenylalanine within region VII (Allende & Allende, 1995). To investigate whether these variant amino acid residues might be responsible for effective GTP utilization, Jakobi and Traugh (1995) mutated both of these residues back to the consensus amino acids. Their results indicated that both single mutants of CK2α and the double mutant CK2α could still use GTP as a phosphate donor. The single and double mutations only altered the relative affinities for ATP and GTP. This finding indicated that at least one other amino acid residue must be responsible for the effective utilization of GTP by CK2. The same authors studied the above-mentioned mutants with respect to the catalytic activity of the reconstructed holoenzyme. The relatively lower affinity for GTP of the holenzyme reconstructed from the mutated CK2α was caused by changes in both the K_m and V_{max} values for GTP and ATP, while for the catalytic subunits, it was a result of changes in the K_m values only. These studies showed that the unique property of the effective utilization of GTP by CK2 was correlated with stimulation of the activity by the regulatory subunits and with the ability to undergo a conformational change upon formation of the holoenzyme.

Srinivasan and collaborators (1999) showed that the dual specificity of CK2 probably originated from the loop situated around the stretch $H_{115}VNNTD_{120}$ in CKα. In their work, they combined site-directed mutagenesis of CK2α with comparative 3D-structure modeling.

Due to significant amino acid sequence similarity (69,5%), kinase CDK2 was chosen to be a good comparative model for CK2α. Based on modeling, a ΔN118 CK2α mutant was constructed. The kinase assay showed decreased affinity of this protein to GTP, in comparison to the wild-type CK2α. The K_m values were 146 and 37 μM, respectively. The results obtained clearly indicate that the adenine/guanine binding region (His115–Asp120) is responsible for the dual specificity of kinase towards phosphate donors (Srinivasan et al., 1999).

The latter study was extended by Jakob and collaborators (2000), who created several mutants of *Xenopus laevis* CK2α with substitutions at positions 118 and 129. They tested them for cosubstrate specificity after their combination with CK2β. The region containing Asn118, known to participate in the recognition of the guanine base, is a part of the sequence $N_{117}NTD_{120}$. This sequence closely resembles the conserved sequence NKXD that is present in G proteins and other GTPases. The study demonstrated that both the CK2α ΔN118 and CK2α N118E mutants produced a 5 to 6-fold increase in the K_m for GTP with little effect on the affinity for ATP.

The mutagenesis by Yde and collaborators (2005) resulted in the first stable and fully active mutant of the human catalytic subunit of protein kinase CK2 that is devoid of dual cosubstrate specificity. The resulting mutant hsCK2α1-335 (human CK2 deleted for the last 56 amino acids) V66A/M163L was designed on the basis of several structures of the enzyme from *Zea mays* in a complex with various ATP-competitive ligands. As structural research revealed the existence of a purine base-binding plane harboring the purine base of ATP and GTP. This plane is flanked in human CK2α by two side-chains of Val66 and Met163, and it adopts a significantly different orientation than it does in other kinase homologues. By exchanging these two flanking amino acids, the cosubstrate specificity is shifted towards strongly favoring ATP. These findings demonstrated that CK2α possesses a sophisticated structural adaptation that favors dual-cosubstrate specificity, a property that may have biological significance.

The mutagenesis studies also provided much insight into the significance of the sequence of the catalytic domain with respect to the CK2α/CK2β interaction. It was reported that CK2α V66A and V66A/W176F were able to interact with CK2β, but this interaction failed to stimulate catalytic activity on the peptide substrate. These results were in contrast to the result with the wild-type α subunit, which was stimulated 4-fold. Nevertheless, the stimulatory response to the cationic modulatory compounds, spermine and polylysine, was the same for holoenzymes reconstituted with the wild-type subunit and all three above-mentioned mutants of the α subunit. The results showed that there must be at least two different interactions between the catalytic α and regulatory β subunit: one that is responsible for stimulation by the β subunit itself and another for mediating the stimulation by polycationic compounds (Jakobi & Traugh, 1992). However, experiments using calmodulin as a substrate for phosphorylation revealed that the insensitivity of the CK2α mutant V66A to CK2β was only apparent. Down-regulation of calmodulin phosphorylation by the CK2β subunit is even enhanced by the V66A mutant. This observation indicated a possible indirect role for Val66 in conferring to the α-subunit a conformation less sensitive to down-regulation (Sarno et al. 1997a).

It is known that the hydrophobic and polar residues of domain II and VII are responsible for the selectivity of a number of specific, potent CK2 ATP-competitive inhibitors, like TBBz (tetrabromobenzimidazole) and TBBt (tetrabromobenzotriazole) (Sarno et al., 2005a). The importance of the same key residues in the hydrophobic portion of the binding site was corroborated by mutational analysis of residues of the human CK2α. Their side chains contribute to the reduction in the internal size of the hydrophobic pocket adjacent to the ATP/GTP-binding site in CK2 (Battistutta et al., 2001; Sarno et al, 2005). Three of these residues (Val66 or Ile66, Ile174, and Met163) are specific to CK2. They are generally replaced by smaller ones in other protein kinases. Both single and double mutants with substitutions for Val66 and Ile174 gave rise to catalytically active CK2α with altered susceptibility to various inhibitors. However, replacement of Met163 by glycine produced a catalytically inactive mutant (Sarno et al., 2005b). Similar data were obtained with yeast CK2α. Mutants with alterations to V67 and I213 (analogous to V66 and I174 of human CK2α) displayed considerably higher K_i values toward inhibitors TBBz and TBBt and only a slight change in the affinity for ATP (Sajnaga et al, 2008). The structural basis for decreased emodin binding to human CK2α resulting from a single point mutation (V66A) has been examined by molecular dynamics (MD) simulations and energy analysis (Zhang & Zhong, 2010). It was found that the V66A mutation resulted in a packing defect due to a change in hydrophobicity. It led to abnormal behavior of the glycine-rich loop, α-helix, and C-loop. The critical role of Ile66 in cosubstrate binding and selection, besides forcing the nucleotide ligands to adopt different positions in the binding pocket, was also demonstrated in a mutational study (Jakobi et al., 1994; Jakobi & Traugh, 1992, 1995).

Chaillot and collaborators (2000) studied the role of Gly177 in conserved region VII of the catalytic domain, which is close to the active site. It was revealed that the CK2α G177K mutant exhibited improved catalytic efficiency for acid peptidic substrates, probably by establishing interactions with the acidic residues.

The acidic residue Asp or Glu of the catalytic loop (corresponding to Glu170 in PKA and conserved in most Ser/Thr protein kinases) is responsible for the binding of basic residues that specify the protein/peptide substrates. In CK2, the residue is replaced by a histidine (His160). Such a substitution could explain the acidophilic properties of CK2, in contrast to the basophilic properties of PKA and other Ser/Thr kinases. The actual role of the His160 in the determination of the site specificity of CK2 was assessed by Dobrowolska and collaborators (1994). Interestingly, subsequent mutational studies in which His160 was replaced with alanine or aspartic acid ruled out any significant role of this residue in substrate recognition (Sarno et al., 1997b).

A CK2α inactive mutant (D156A) was produced based on structural homology to kinase PKA. The mutant protein was able to compete efficiently with the wild-type CK2α for the regulatory β subunits. Although it does not exhibit kinase activity, the D156A mutant can bind CK2β to form an inactive holoenzyme. Moreover, the mutant abolishes the inhibitory effect of CK2β on CK2α-mediated phosphorylation of calmodulin. These results suggest that CK2α D156A may be a useful dominant-negative mutant for elucidation of the cellular functions of the CK2 regulatory subunit (Cosmelli et al., 1997).

2.1.4 Mutations of CK2α in the glycine-rich loop

The glycine-rich sequence (G-loop) is one of the most critical structures of protein kinases, since it contributes in many ways to enzyme activity. This multifunctional structural element participates in nucleotide binding, substrate recognition, catalysis, and regulation of activity (Bossemeyer et al., 1994). In their extensive mutational studies combined with biochemical characterization, Sarno and collaborators (1999) confirmed that some basic residues in the glycine-rich loop of the CK2α, particularly Lys49, are implicated in substrate recognition and inhibition by polyanions. Another residue located within this region, Gly48, is involved in binding the ATP phosphate moiety. Replacement of Gly48 by alanine in CK2α affected its catalytic efficiency and specificity. It is thought that alanine causes this phenotype by creating an electrostatic barrier between ATP and the peptide substrate (Chaillot et al., 2000).

2.1.5 Mutations of CK2α in the C-terminal region

The C-terminal region of vertebrate CK2α is composed of 54 amino acids. Knowledge of this segment is rather poor, except for phosphorylation by kinase p34[Cdc2] and interaction with isomerase Pin1 (Bosc et al., 1995; Messenger et al., 2002). It is known from the publications on crystallization of CK2 that the catalytic subunits are particularly sensitive to degradation, which makes the crystallization process of the entire subunit difficult (Niefind et al., 2000, 2001). Truncation at the C-terminus reduced the intrinsic degradability of CK2α and allowed its crystallization and the determination of its 3D structure. Starting from sequence alignments of C-termini from different CK2αs, Grasselli and collaborators (2004) constructed a mutant carrying the substitution of two distal prolines with alanines (P382A/P384A). Most intriguing was the resistance of the mutant to proteolytic degradation, which makes this protein an excellent candidate for crystallization of the entire CK2α subunit.

Bischoff and collaborators (2011) have recently determined for the first time the structure of the full-length human CK2α`[C336S] subunit. A point mutation of CK2α` was necessary to prevent covalent dimerization from intermolecular disulfide bridges formed by Cys336. However, these results shed light on the differences between the two catalytic subunits, α and α` (e.g., significantly lower affinity of CKα` towards CK2β relative to that of CK2α).

2.1.6 Mutagenesis of CK2α in other regions

Determination of the structure of the CK2 holoenzyme and individual subunits provided knowledge about the nature and location of the interface between catalytic and regulatory subunits (Niefind et al., 2001). Using structure-guided alanine-scanning mutagenesis combined with isothermal titration calorimetry (ITC), energetic "hot spots" were identified on the surface of CK2α that determine the α/β subunit interaction (Raaf et al., 2011). Three single and one double CK2α subunit mutants were produced, in which individual hydrophobic amino acids located within the CK2α interface were replaced by alanine. The ITC analysis of CK2α mutants revealed that substitution of Leu41 and Phe54 were most disruptive to binding of CK2β. Moreover, the L41A and F54A mutants retained their kinase activity, compared to the wild-type CK2. Based on the results mentioned above, it can be claimed that these residues are suspected of being interaction "hot spots" (Raaf et al., 2011).

The amino-acid sequence and the structure of yeast protein kinase CK2α differ from those of CK2α' and other eukaryotic CK2α subunits. CK2α is unique in containing a 38-amino-acid loop consisting of two α-helical structures situated close to structures engaged in ATP/GTP and substrate binding (Niefind et al., 2001). Modeling of the tertiary structure of the CK2α showed that, after removing both α-helical motifs, the CK2α subunit assumes a structure that is more similar to that of CK2α′ than it is to the structure of intact CK2α. The deletion of the 38 amino acids from CK2α drastically decreases its catalytic efficiency. Its characteristics are similar to yeast CK2α′ with respect to sensitivity to salt, heparin and spermine (Sajnaga et al., 2008) (Fig. 3).

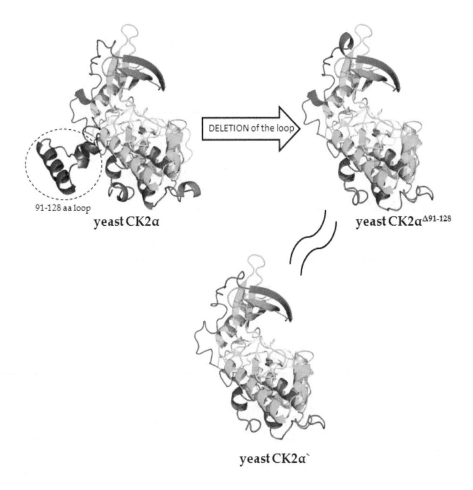

Fig. 3. Conformational consequences of mutagenesis of the yeast CK2α catalytic subunit.

CK2 residues	Location	Mutant	Reference/source
Substrate recognition and inhibition by polyanions			
K49	Subd. I; Gly loop	K49A	Sarno et al., 1999
K74			
K75	Subd. II/III, Lys rich loop	K74-77A, K77A	Sarno et al., 1997a, 1998, 1999, Vaglio et al. 1996
K76			Gatica et al., 1994[1]
K77			
K79		K79A,	
R80	Subd. III, Helix C	R80A/K83A	Sarno et al., 1998, 1999
K83			
K122	Subd.V, Linker region	K122A	Sarno et al., 1997a, 1999
H160	Subd. VIb, Catalytic loop	H160D	Dobrowolska et al., 1994
R191			
R195		R191, 195, K190A	Sarno et al., 1997a, 1998,
K190	Subd. VIII, p+1 loop	K198A	1999; Vaglio et al., 1996
K198			
Catalytic efficiency and specificity			
G48[2]	Subd. I, Gly loop	G48D	Chaillot et al., 2000
V66	Subd. II	V66A M163A	Yde et al., 2005
M163	Subd. VIb	CK2α 1-335	

V66	Subd. II	V66A/I174A	Sarno et al., 2005
I174	Subd. VII		

V66/W176	Subd. II, Subd. VII	V66A/W176F	Jakobi and Traugh, 1992, 1995

N118	Subd. V, ATP/GTP binding region	N118A, CK2α$^{\Delta N118}$	Srinivasan et al., 1999; Jakob et al., 2000

D156	Subd. VIb	D156A	Cosmelli et al., 1997

M163	Subd. VIb	M163G	Sarno et al., 2005

G177	Subd. VII	G117K	Chaillot et al., 2000

N189	Subd. VII, Activating segment	N189R	Srinivasan et al., 1999

Regulation by β subunit			

L41	Subd. I	L41A	Raaf et al., 2011

L54	Subd. I, ATP/GTP binding region	L54A	Raaf et al., 2011

V66	Subd. II	V66A,	Sarno et al., 1997b

W176	Subd. VII	V66A/W176F	Jakobi &Traugh, 1992

Constitutive activity			

M6 –V30	N-terminus	Δ2-12, Δ2-18, Δ2-24, Δ2-30	Sarno et al., 2001, 2002; Cristiani et al., 2011

Y125	Subd. V, Hinge region	Y125F[4]	Cristiani et al., 2011

E180	Subd. VII, Activation segment	E180A	Sarno et al., 2002

E182	Subd. VII, Activation segment	Y182F	Sarno et al., 2002
Stability			
M336-Q393	C-terminus	Δ336-393	Ermakova et al., 2003
P382		P382A	
	C-terminus		Grasselini et al., 2004
P384		P384A	
C336[3]	C-terminus	C336S	Bischoff et al., 2011

[a]The residue numbers correspond with those of human CK2α, unless otherwise indicated. The Roman numerals indicate the eleven conserved subdomains present in the catalytic domain of all protein kinases (Hanks & Hunter, 1995). Abbreviations: [1]CK2α from *Xenopus laevis;* [2]CK2α from *Yarrovia lipolytica;* [3]Human CK2α'; [4]*in silico* mutation.

Table 1. Summary of CK2α mutants[a]

The deletion of the loop of amino acids 91-128 from yeast CK2α led to behavioral and structural similarity to CK2α` (Sajnaga et al., 2008). The 3D models of proteins were created using the SWISS-MODEL software based on protein structure templates (PDB code 1ds5D) available in the Protein Data Bank and visualized with the PyMOL software.

Chimeras of different kinases can be easily engineered using recombinant DNA technology and used in studies on the structure and function of kinase. To study the effect of CK2β on the activity of CK1α, Jedlicki and collaborators (2008) generated CK2α/CK1α chimeras that were able to bind tightly to the CK2β regulatory subunit, but maintain the peptide substrate specificity of CK1. This is related to the capacity of the CK2β to regulate the activity of CK2α, as well as other protein kinases, such as A-Raf, C-Mos, and Chk1. It has been shown that a chimera combining a large part of the CK1α kinase with the N-terminal region of CK2α that is responsible for binding CK2β can be stimulated by this subunit. It is possible that such chimeras could be used to test the presence of "the docking site" on the CK2β subunit, which would bring substrate molecules near the catalytic subunits.

2.2 Mutagenesis of the regulatory subunit CK2β

From the primary sequence of the β subunit, it is obvious that the charged amino acids are not equally distributed. The acidic residues are clustered in the N-terminal half, whereas the basic residues are clustered in the C-terminal part of the molecule. Mutational studies have shown that, in contrast to cyclins, which invariably act as indispensable activators of CK2-related CDKs, the CK2β subunit fulfills antagonist functions. The features of CK2β can be explored by generating large synthetic fragments, some of which reproduce the C-terminal moiety and thus stimulate its catalytic activity. Fragments reproducing segments of the N-terminal sequence are inhibitory, which becomes especially evident when calmodulin is the substrate (Marin et al, 1992, 1995; Meggio et al, 1994; Sarno et al, 1997a).

2.2.1 Mutations of CK2β that affect autophosphorylation

The CK2β subunit is known to be autophosphorylated by the catalytic subunit. Autophosphorylation occurs on serine residues at positions 2 and 3 in the amino-terminal region of the molecule. Both these serines fit CK2 consensus specificity requirements (Marin et al, 1992). This finding was corroborated by the fact that the mutant S2,3G (*i.e.*, S2G/S3G) is completely incapable of autophosphorylation (Hinrichs et al, 1993). Deletion of the first four amino acids (CK2β Δ1-4), which eliminated autophosphorylation of CK2β, had no significant effect on the reconstruction of CK2 holoenzymes nor on their catalytic activity, thermostability, and responsiveness to polylysine. Unlike the wild-type CK2β, however, CK2β Δ1-4 failed to confer to the reconstituted holoenzyme the typical responsiveness to NaCl stimulation. These results indicated that autophosphorylation sites are not required on CK2 for conferring a stable structure and full catalytic activity. In contrast an autophosphorylation site is implicated in the NaCl-dependent fine-tuning of CK2 activity (Meggio et al., 1993). Interestingly, the acidic stretch heavily influences autophosphorylation of the β subunit, even though Ser2 is more than 50 amino acids away in the primary sequence (Boldyreff et al., 1994).

2.2.2 Mutations of CK2β that affect binding with CK2α

In order to shed light on the mechanisms by which the CK2β subunits affect the catalytic properties of CK2 and to elucidate the molecular interactions between the catalytic and regulatory subunits of CK2, Boldyreff and collaborators (1992, 1993) generated a number of mutants of the CK2β subunit, which were tested for their ability to functionally replace the wild-type CK2β. These authors showed that deletion of the last 44 residues of the C-end (CK2β Δ171-215) eliminated the capacity to form tetramers with CK2α and to stimulate activity. However, deletion of the last 34 amino acids (CK2βΔ181-215) yielded an active CK2β that had lower affinity for CK2α. Shorter deletions (*e.g.*, CK2β Δ194-215) did not affect the interaction between the catalytic and regulatory subunits of CK2. Boldyreff and collaborators demonstrated that deletion mutants in which the last 45 or more amino acids are missing were not able to assemble with the α subunit. These data identified the C-terminal segment of CK2β as essential for association with the CK2α subunit, with special reference to its 171-180 stretch, which is indispensable both to form tetrameric CK2 and to stimulate activity of the CK2α catalytic subunit (Boldyreff et al., 1994). Tight interaction between the CK2α and CK2β subunits, accomplished by the C-terminal part of the CK2β subunit, was also described (Kusk et al., 1995; Marin et al., 1997).

Mutagenesis along with crosslinking and peptide studies have shown that the acidic amino acid stretch of CK2β from residues 55-64 interacts with a corresponding basic stretch of the CK2α subunit. However, these weak electrostatic interactions seem to determine the activity of, but not the formation of, the CK2 holoenzyme (Krehan et al., 1996, Sarno et al, 1997b).

Kusk and collaborators (1995) used mutagenesis of CK2 subunits with a yeast two-hybrid system to explore domains involved in intersubunit contact. [In the yeast two-hybrid system, a peptide or protein is fused to part A of a transcriptional activator. Another peptide or protein is fused to part B. Transcriptional activation of an easily assayed reporter gene occurs only when part A and part B come together. Parts A and B themselves cannot interact to form the transcriptional activator, nor can either part individually (part A, the part A fusions, part B, and the part B fusions) cause the reporter to be expressed. However, if the fusions interact, part A and part B can come together, and the reporter is activated. This is an indication that the peptides or proteins in the fusions can interact.] A series of plasmid constructs was prepared. They encoded N-terminal or C-terminal truncations of the CK2α and CK2β subunits to indicate which regions of the subunits were engaged in CK2 holoenzyme formation in yeast cells. The data revealed that the regulatory CK2β subunit has a modular structure. An N-terminal domain (residues 20-145) is responsible for homodimerization (CK2β/CK2β). A C-terminal domain (residues 152-200) is necessary for heterodimerization (CK2α/CK2β). Amino acid residues 1 to 20 in the N-terminus and 351 to 391 in the C-terminus of CK2α are dispensable for interaction with the regulatory subunit.

2.2.3 Mutations of CK2β that affect the activity of CK2α

The modulation of CK2α subunit activity by CK2β has a stimulatory effect on most substrates. However, when calmodulin is used as the substrate, the CK2β subunit almost completely inhibits the activity of the catalytic subunit (Guerra et al., 1999). This inhibition can be overcome by addition of polylysine (Meggio et al, 1992). Mutagenesis studies on the CK2β subunit revealed an acidic stretch (amino acids 55-64) that is responsible for the inhibitory effect and for the stimulation by polylysine (Meggio et al., 1994). Interestingly, mutants of CK2β bearing substitutions at positions 55, 57, and 59-64 to alanine produced up to 4-fold more active holoenzyme after assembling with the catalytic α subunit than did the wild type. At the same time, these mutants were refractory to the stimulatory effect of polylysine. This finding revealed that the acidic N-terminal cluster of CK2β, especially Asp55 and Glu57, is involved in intrinsic down-regulation of CK2 basal activity and has been implicated in responsiveness to various effectors (Boldyreff et al., 1993, 1994).

Other data provided by Hinrichs and collaborators (1995) demonstrated that Pro58 located in the center of the acidic segment also constitutes an important structural feature affecting the function of down-regulation of CK2β towards the catalytic subunits. The effect of a mutation of proline to alanine resulted in an effect that was similar to mutation of the acidic residues alone. It produced hyperactive CK2β subunits that stimulated the CK2α activity to a greater extent than did the wild-type CK2β subunit.

2.2.4 Mutations of CK2β that affect export of the holoenzyme

It is known that protein kinase CK2 is present in not only the cytoplasm, nuclei, and several other cell organelles, but also on the external side of the cellular membrane (Kubler et al, 1983). Rodrigez and collaborators (2008) have studied the role of CK2β in the export of the holoenzyme to the extracellular membrane through deletion and point mutations. The region of CK2β between amino acids 20 and 33 was found to be necessary, but not sufficient, to allow the catalytic subunits to function as an ectokinase. An important function of this region is fulfilled by Phe21 and Phe22, which anchor the loop of the 20-33 sequence. Another key element of this region is constituted by the acidic residues in positions 26-28. They are exposed to the medium, free to interact with other proteins (Bolanos-Garcia et al, 2006).

2.2.5 Mutation of CK2β that affects its stability

Overexpression of CK2 catalytic subunits leads to increased cell proliferation and transformation, while overexpression of the regulatory CK2 subunit is associated with decreased proliferation in yeast and mammalian cells (Li et al., 1999; Lebrin et al., 2001; Vilk et al., 2001). Moreover, CK2β is physiologically expressed at a higher level than CK2α, and the excess of the regulatory subunit is rapidly ubiquitinated and degraded in a proteasome-dependent manner (Luscher & Litchfield, 1994; Zhang et al., 2002). To protect CKβ from the degradation machinery and to stabilize it, six surface-exposed lysine residues were mutated to arginine (French et al., 2007). The 6KR mutant functioned as normal CK2β, but it was not sensitive to proteasome inhibition. The physiological role of mutagenesis-mediated CK2β stabilization was also examined with the use of cell proliferation assays. A significant decrease in proliferation was observed in cells expressing the 6KR mutant when compared to wild-type CK2β. The authors suggest that the stabilized form of the CK2 regulatory subunits can be utilized to inhibit cell proliferation in cancer cells (French et al., 2007).

2.3 Mutagenesis of CK2 substrates

Protein kinase CK2 is a multi-substrate enzyme with a large number of cellular partners. In 2003, Meggio and Pinna updated the list of 307 CK2 substrates with 308 sites phosphorylated by CK2 (Meggio & Pinna, 2003). This number is now out-of-date, as novel CK2 protein substrates are discovered every year. A *bona fide* CK2 substrate may possess one or several phosphoacceptor sites affected by CK2, but an analysis of the initial amino acid sequences of possible CK2 partners may show a dozen or so putative CK2 sites. Site-directed mutagenesis is a useful tool to create CK2 substrate mutants. Such proteins are produced (1) to indicate precisely the phosphorylatable amino acid, (2) to study the physiological significance of CK2-mediated phosphorylation of a given protein substrate, or (3) to confirm the physiological relevance of CK2-mediated phosphorylation. Presented below are several examples of the mutagenesis of CK2 substrates.

Mdm2 is a cellular oncoprotein that down-regulates the growth suppressor protein p53 (Barak et al., 1992). Computer analysis of the amino acid sequence of Mdm2 revealed 19 putative CK2 phosphorylation sites. Three Mdm2 mutants with deletions at codons 1-114, 93-285, and 271-491 were produced to exclude sites that are not affected by CK2. The phoshorylation assays revealed that only the central part of Mdm2 is phosphorylated. Based on further detailed analysis of the remaining CK2 consensus sites, Ser269 was chosen to be the most promising. Using overlap extension PCR (see section 2.7 in the chapter by Sturtevant), the Mdm2 point mutant S269A was constructed and the relevant CK2 phosphorylation site was finally discovered (Götz et al., 1999).

In some protein substrates, putative CK2 phosphorylation sites are located close to one another, and thus several point mutants had to be produced to score them. The consensus sequence analysis of the N-terminal domain of the human transcription factor Tcf-4 indicated multiple sites that fit the motif for CK2 phosphorylation. No CK2-mediated phosphorylation was detected on the Tcf-4 fragments comprising amino acids 1-30 and 1-49. Thus, the best candidates for CK2-affected amino acids were the serine residues located in the Tcf-4 peptide $T_{54}NQDSSSDSEAERRP_{68}$. Three Tcf-4 mutants, one triple point mutant (S58A/S59A/S60A) and two single point mutants (S58E and S60E) were made to help indicate the phosphorylatable amino acid. *In vitro* phosphorylation assays revealed that all three adjacent serines are modified by CK2 with different efficiencies (Miravet et al., 2002).

Sic1 is a yeast protein that specifically inhibits Clb/Cdk activity in the G1 phase, so that DNA replication is suppressed (Verma et al., 2001; Nash et al., 2001). Moreover, Sic1 undergoes multistep phosphorylation. Therefore, Sic1 phosphorylation occurs at several positions. One looks like the CK2 consensus site. CK2-mediated phosphorylation of Sic1 within the Q_{199}ESEDEED sequence was confirmed both *in vitro* and *in vivo* in *Saccharomyces cerevisiae* cells (Coccetti et al., 2004, 2006). Mutations of the CK2 consensus site on Sic1 (S201A and S201E) alter the coordination between cell growth and division. They also change the level and time-course of S-Cdk kinase activity. These mutation data strongly support the physiological relevance of Sic1 phosphorylation for inhibitory activity (Coccetti et al., 2004).

The regulatory effect of CK2 activity on the Wnt signaling pathway is widely known (Pinna, 2002; Litchfield, 2003). Kinase phosphorylates and interacts with β-catenin and thus enhances the stability and transcriptional activity of β-catenin (Song et al, 2003; Seldin et al, 2005). The AKT/PKB kinase is also a well-known CK2 substrate and interacting partner. CK2-mediated phosphorylation at Ser129 causes AKT hyperactivation (Di Maira et al, 2005; Guerra, 2006). CK2 may link the two pathways..

To elucidate the roles of CK2 in the Wnt and AKT/PKB signaling pathways, the AKT phosphorylation-deficient mutant (S129A) was overexpressed in an embryonic cell line. The β-catenin-dependent transcriptional activity was analyzed. The data obtained indicate that blockage of AKT phosphorylation by CK2 impairs β-catenin activity and decreases its stability. Therefore, CK2-mediated AKT phosphorylation at Ser129 is a necessary step in the up-regulation of the β-catenin transcriptional activity in human embryonic kidney cells (Ponce et al., 2011).

Besides phosphorylation of numerous cellular proteins, CK2 directly interacts with many of them forming protein-protein complexes (Litchfield, 2003). Both catalytic and regulatory CK2 subunits can interact with different proteins, independently of the holoenzyme (Bibby et al., 2005). Wee1 kinase, involved in cell cycle progression, is one such CK2 protein partner. The Wee1 kinase is a key inhibitor of cyclin-dependent kinase (CDK1) and mitotic entry in eukaryotes. Several deletion mutants of the Wee1 catalytic domain were produced to investigate the interaction with CK2 subunits. Immunoprecipitation experiments revealed that Wee1 binds CK2β via two domains of Wee1 (comprising amino acids 59-71 and 232-332) and two regions of CK2β (comprising residues 1-5 and 155-170). Although the interaction does not affect Wee1 activity, it up-regulates CDK1 by reversing the Wee1-mediated inhibitory effect on CDK1. These findings reinforce the notion that CK2β can serve other protein kinases. It may be a universal regulatory subunit that can act independently of the CK2 holoenzyme (Olsen et al., 2010).

3. Conclusion

Even 58 years after its first description (Burnett & Kennedy, 1954), the story of protein kinase CK2 has not been fully clarified. This enzyme catalyzes phosphorylation of over 300 substrates. They are characterized by having multiple acidic residues surrounding the phospho-acceptor amino acid. Consequently, CK2 plays a key role in several physiological and pathological processes (Guerra & Issinger, 2008). After all those years of research, we are still asking the question: how is it possible that one kinase can be involved in so many different biochemical processes in the cell? Using different biochemical and genetic methods, we have solved several problems connected with the structure and mechanism of the catalytic action of this enigmatic protein kinase. The application of mutagenesis methods in many cases has helped us and will continue to help us get answers to many problems connected with CK2 activity. Among them are the following:

- The interaction between subunits
- Catalytic specificity and efficiency
- Substrate recognition
- Regulation by the β-subunit
- Stability of the subunits
- Interactions with modulators and substrates
- The effect of phosphorylation on catalytic activity
- Constitutive CK2 activity.

A protein kinase, such as CK2, is difficult to explore with respect to its physiological functions. CK2 has been shown to be involved in numerous aspects of cell proliferation and survival, including cell cycle progression and apoptosis control (Ahmad et al., 2008; Ahmed et al., 2002; Batistuta, 2009; Gyenis & Litchfield, 2008; Meggio & Pinna, 2003; Litchfield, 2003). Alterations in the levels or activity of CK2 have been implicated in a variety of human diseases, including cancers (Guerra & Issinger, 2008). All these observations raise important questions regarding the mechanisms that control CK2 activity and specificity. These questions have a special value, since defects in regulation of these processes could contribute to tumorigenesis.

In this context, the application of mutagenesis methods, together with other techniques (*e.g.*, molecular modeling), may be very useful in designing highly effective and specific inhibitors that are promising for CK2-based target therapy.

4. Acknowledgement

The 3D protein structure models of CK2 were kindly constructed by Maciej Masłyk, PhD. (Department of Molecular Biology, Institute of Biotechnology, The John Paul II Catholic University of Lublin, Poland)

5. References

Ahmad K.A., Wang G., Unger G., Slaton J., Ahmed K. (2008) Protein kinase CK2 – A key suppressor of apoptosis. *Advances in Enzyme Regulation*, Vol. 48, No. 1, (April 2008), pp. 179-187, ISSN 0065-2571.

Ahmed, K., Gerber, D.A., Cochet, C. (2002) Joining the cell survival squad: an emerging role for protein kinase CK2. Trends in Cellular Biology, Vol. 12, No. 5, (May 2002), pp. 226-230, ISSN 0962-8924.

Adler, V., Pincus, M.R., Minamoto, T., Fuchs, S. Y., Bluth, M.J., Brandt-Rauf, P.W., Friedman, F.K., Robinson, R.C., Chen, J.M., Wang, X.W., Harris, C.C. & Ronai, Z. (1997). Conformation-dependent phosphorylation of p53. *Proceedings of the National Academy of Sciences of the United States of America*, Vol. 94, No. 5, (March 1997), pp. 1686-1691, ISSN 0027-8424.

Allende, J. E. & Allende, C.C. (1995). Protein kinases. 4. Protein kinase CK2: an enzyme with multiple substrates and a puzzling regulation. *The FASEB Journal*, Vol. 9, No. 5, (March 1995), pp. 313-323, ISSN 0892-6638.

Arnold K., Bordoli L., Kopp J., Schwede T. (2006) The SWISS-MODEL workspace: a web-based environment for protein structure homology modelling. *Bioinformatics* Vol.22, No. 2, (January 2006), pp. 195-201, ISSN 1367-4803

Barak, Y. & Oren, M. (1992). Enhanced binding of a 95 kDa protein to p53 in cells undergoing p53-mediated growth arrest. *The EMBO Journal*, Vol. 11, No. 6, (June 1992), pp. 2115-2121, ISSN 0261-4189.

Battistutta, R. (2009) Protein kinase CK2 in health and disease: Structural bases of protein kinase CK2 inhibition. *Cellular and Molecular Life Sciences*, Vol. 66, No. 11-12, (June 2009), pp. 1868-1889, ISSN 1420-682X.

Battistutta, R., Sarno, S., De Moliner, E., Marin, O., Issinger, O.-G., Zanotti, G., Pinna, L.A. (2000)_The crystal structure of the complex of *Zea mays* alpha subunit with a fragment of human beta subunit provides the clue to the architecture of protein kinase CK2 holoenzyme. *European Journal of Biochemistry* 267, No. 16, (August 2000), pp. 5184-5190, ISSN 0014-2956.

Battistutta, R., De Moliner, E., Sarno, S., Zanotti, G. & Pinna, L.A. (2001). Structural features underlying selective inhibition of protein kinase CK2 by ATP site-directed tetrabromo-2-benzotriazole. *Protein Science*, Vol. 10, No. 11, (November 2001), pp. 2200-2206, ISSN 0961-8368

Bibby, A.C. & Litchfield, D.W. (2005). The multiple personalities of the regulatory subunit of protein kinase CK2: CK2 dependent and CK2 independent roles reveal a secret identity for CK2beta. *International Journal of Biological Sciences*, Vol. 1, No. 2, (April 2005), pp. 67-79, ISSN 1449-2288

Bischoff, N., Olsen, B., Raaf, J., Bretner, M., Issinger, O.-G. & Niefind, K. (2011). Structure of the human protein kinase CK2 catalytic subunit CK2alpha' and interaction thermodynamics with the regulatory subunit CK2beta. *Journal of Molecular Biology*, Vol. 407, No. 1, (March 2011), pp. 1-12, ISSN 1089-8638.

Becker, W., Weber, Y., Wetzel, K., Eirmbter, K., Tejedor, F.J., Joost, H.-G. (1998) Sequence characteristics, subcellular localization, and substrate specificity of DYRK-related kinases, a novel family of dual specificity protein kinases. *The Journal of Biological Chemistry*, Vol. 273, No. 40, (October 1998), pp. 25893-25902, ISSN 0021-9258.

Biondi R.M. and Nebreda A.R. (2003) Signalling specificity of Ser/Thr protein kinases through docking-site-mediated interactions. *Biochemical Journal*, Vol. 372, Pt. 1, (May 2003), pp. 1-13, ISSN 0264-6021.

Bolanos-Garcia, V.M., Fernandez-Recio, J., Allende, J.E. & Blundell, T.L. (2006). Identifying interaction motifs in CK2beta--a ubiquitous kinase regulatory subunit. *Trends in Biochemical Sciences*, Vol. 31, No. 12, (December 2006), pp. 654-661, ISSN 0968-0004.

Boldyreff, B., Meggio, F., Pinna, L.A. & Issinger, O.-G. (1992). Casein kinase-2 structure-function relationship: creation of a set of mutants of the beta subunit that variably surrogate the wildtype beta subunit function. *Biochemical and Biophysical Research Communications*, Vol. 188, No. 1, (October 1992), pp. 228-234, ISSN 0006-291X

Boldyreff, B., Meggio, F., Pinna, L.A. & Issinger, O.-G. (1993). Reconstitution of normal and hyperactivated forms of casein kinase-2 by variably mutated beta-subunits. *Biochemistry*, Vol. 32, No. 47, (November 1993), pp. 12672-12677, ISSN 0006-2960

Boldyreff, B., Meggio, F., Pinna, L.A. & Issinger, O.-G. (1994). Protein kinase CK2 structure-function relationship: effects of the beta subunit on reconstitution and activity. *Cellular & Molecular Biology Research*, Vol. 40, No. 5-6, (January 1994), pp. 391-399, ISSN 0968-8773

Boldyreff, B., James, P., Staudenmann, W., Issinger, O.-G. (1993) Ser2 is the autophosphorylation site in the beta subunit from bicistronically expressed human casein kinase-2 and from native rat liver casein kinase-2 beta. *European Journal of Biochemistry*, Vol. 218, No. 2, (December 1), pp. 515-521, ISSN 0014-2956.

Boldyreff, B., Mietens, U., Issinger O.-G. (1996) Structure of protein kinase CK2: dimerization of the human beta-subunit. *FEBS Letters*, Vol. 379, No.2, (January 1996), pp. 153-156, ISSN 0014-5793.

Brognard, J. & Hunter, T. (2011) Protein kinase signaling networks in cancer. *Current Opinion in Genetics and Development*, Vol. 21, No.1 , (February 2011), pp. 4-11, ISSN 0959-437X.

Bosc, D. G., Slominski, E., Sichler, C. & Litchfield, D. W. (1995). Phosphorylation of casein kinase II by p34cdc2. Identification of phosphorylation sites using phosphorylation site mutants in vitro. *The Journal of Biological Chemistry*, Vol. 270, No. 43, (October 1995), pp. 25872-25878, ISSN 0021-9258

Bossemeyer, D. (1994). The glycine-rich sequence of protein kinases: a multifunctional element. *Trends in Biochemical Sciences*, Vol. 19, No. 5, (May 1994), pp. 201-205, ISSN 0968-0004

Burnett, G. & Kennedy E.P. (1954) The enzymatic phosphorylation of proteins. *The Journal of Biological Chemistry*, Vol. 211, No. 2, (December 1954), pp. 969-980, ISSN 0021-9258.

Canton, D.A., Zhang, C., Litchfield, D.W. (2001) Assembly of protein kinase CK2: investigation of complex formation between catalytic and regulatory subunits using a zinc-finger-deficient mutant of CK2beta. *Biochemical Journal*, Vol. 358, Pt. 1, (August 2001), pp. 87-94, ISSN 0264-6021.

Chantalat, L., Leroy, D., Filhol, O., Nueda, A., Benitez, M.J., Chambaz, E.M., Cochet, C., Dideberg, O. (1999) Crystal structure of the human protein kinase CK2 regulatory subunit reveals its zinc finger-mediated dimerization. *The EMBO Journal*, Vol. 18, No. 11, (June 1999), pp. 2930-2940, ISSN 0261-4189.

Chaillot, D., Declerck, N., Niefind, K., Schomburg, D., Chardot, T. & Meunier, J.C. (2000). Mutation of recombinant catalytic subunit alpha of the protein kinase CK2 that affects catalytic efficiency and specificity. *Protein Engineering*, Vol. 13, No. 4, (April 2000), pp. 291-298, ISSN 0269-2139.

Chen, M., Li D., Krebs, E.G., Cooper, J.A. (1997) The casein kinase II beta subunit binds to Mos and inhibits Mos activity. *Molecular and Cellular Biology*, Vol. 17, No. 4, (April 1997), pp. 1904-1912, ISSN 0270-7306.

Coccetti, P., Rossi, R. L., Sternieri, F., Porro, D., Russo, G. L., di Fonzo, A., Magni, F., Vanoni, M. & Alberghina, L. (2004). Mutations of the CK2 phosphorylation site of Sic1 affect cell size and S-Cdk kinase activity in Saccharomyces cerevisiae. *Molecular Microbiology*, Vol. 51, No. 2, (January 2004), pp. 447-460, ISSN 0950-382X.

Coccetti, P., Zinzalla, V., Tedeschi, G., Russo, G. L., Fantinato, S., Marin, O., Pinna, L. A., Vanoni, M. & Alberghina, L. (2006). Sic1 is phosphorylated by CK2 on Ser201 in budding yeast cells. *Biochemical and Biophysical Research Communications*, Vol. 346, No. 3, (August 2006), pp. 786-793, ISSN 0006-291X.

Cohen, P. (2002) Protein kinases - the major drug targets of the twenty-first century? *Nature Reviews Drug Discovery*, Vol. 1, No. 4, (April 2002), pp. 309-315, ISSN 1474-1776.

Cosmelli, D., Antonelli, M., Allende, C. C. & Allende, J. E. (1997). An inactive mutant of the alpha subunit of protein kinase CK2 that traps the regulatory CK2beta subunit. *FEBS letters*, Vol. 410, No. 2-3, (June 1997), pp. 391-396, ISSN 0014-5793.

Cristiani, A., Costa, G., Cozza, G., Meggio, F., Scapozza, L. & Moro, S. (2011). The role of the N-terminal domain in the regulation of the "constitutively active" conformation of protein kinase CK2alpha: insight from a molecular dynamics investigation. *ChemMedChem*, Vol. 6, No. 7, (July 2011), pp. 1207-1216, ISSN 1860-7187.

Di Maira, G., Salvi, M., Arrigoni, G., Marin, O., Sarno, S., Brustolon, F., Pinna, L. A. & Ruzzene, M. (2005). Protein kinase CK2 phosphorylates and upregulates Akt/PKB. *Cell death and differentiation*, Vol. 12, No. 6, (June 2005), pp. 668-677, ISSN 1350-9047.

Dobrowolska, G., Meggio, F., Marin, O., Lozeman, F. J., Li, D., Pinna, L. A. & Krebs, E. G. (1994). Substrate recognition by casein kinase-II: the role of histidine-160. *FEBS letters*, Vol. 355, No. 3, (December 1994), pp. 237-241, ISSN 0014-5793.

Duncan J.S., Litchfield D.W. (2008) Too much of a good thing: The role of protein kinase CK2 in tumorigenesis and prospects for therapeutic inhibition of CK2. *Biochemica et Biophysica Acta*, Vol. 1784, No. 1, (January 2008), pp. 33-47, ISSN 0006-3002.

Duncan, J.S., Turowec, J.P., Vilk, G., Li, S.S.C., Gloor, G.B., Litchfield, D.W. (2010) Regulation of cell proliferation and survival: Convergence of protein kinases and caspases. *Biochemica et Biophysica Acta*, Vol. 1804, No. 3, (March 2010) pp. 505-510, ISSN 0006-3002.

Ermakova, I., Boldyreff, B., Issinger, O. G. & Niefind, K. (2003). Crystal structure of a C-terminal deletion mutant of human protein kinase CK2 catalytic subunit. *Journal of Molecular Biology*, Vol. 330, No. 5, (July 2003), pp. 925-934, ISSN 0022-2836.

Faust, M., Montenarh, M. (2000) Subcellular localization of protein kinase CK2. A key to its function? Cell and Tissue Research, Vol. 301, No. 3, (September 2000), pp. 329-340, ISSN 0302-766X.

French, A. C., Luscher, B. & Litchfield, D. W. (2007). Development of a stabilized form of the regulatory CK2beta subunit that inhibits cell proliferation. *The Journal of Biological Chemistry*, Vol. 282, No. 40, (October 2007), pp. 29667-29677, ISSN 0021-9258.

Gatica, M., Jedlicki, A., Allende, C. C. & Allende, J. E. (1994). Activity of the E75E76 mutant of the alpha subunit of casein kinase II from Xenopus laevis. *FEBS Letters*, Vol. 339, No. 1-2, (February 1994), pp. 93-96, ISSN 0014-5793.

Ghosh & Adams (2011) Phosphorylation mechanism and structure of serine-arginine protein kinases. *The FEBS Journal*, Vol. 278, No. 4, (February 2011), pp. 587-597, ISSN 1742-4658

Gianoncelli, A., Cozza, G., Orzeszko, A., Meggio, F., Kazimierczuk, Z., Pinna, L.A. (2009) Tetraiodobenzimidazoles are potnt inhibitors of protein kinase CK2. *Bioorganic & Medicinal Chemistry*, Vol. 17, No 20, (October 2009), pp. 7281-7289, ISSN 0968-0896.

Gietz, R.D., Graham, K.C. and Litchfield, D.W. (1995) Interactions between the subunits of casein kinase II. *The Jornal of Biological Chemistry*, Vol. 270, No. 22, (June 1995), pp. 13017–13021. ISSN 0021-9258.

Glover C.V. (1998) On the physiological role of casein kinase II in *Saccharomyces cerevisiae*. *Progress in Nucleic Acid Research & Molecular Biology*, Vol. 59, (May 1998), pp. 95–133, ISSN 0079-6603.

Gotz, C., Kartarius, S., Scholtes, P., Nastainczyk, W. & Montenarh, M. (1999). Identification of a CK2 phosphorylation site in mdm2. *European Journal of biochemistry / FEBS*, Vol. 266, No. 2, (December 1999), pp. 493-501, ISSN 0014-2956

Gotz, C., Scholtes, P., Prowald, A., Schuster, N., Nastainczyk, W. & Montenarh, M. (1999). Protein kinase CK2 interacts with a multi-protein binding domain of p53. *Molecular and Cellular Biochemistry*, Vol. 191, No. 1-2, (January 1999), pp. 111-120, ISSN 0300-8177

Graham, K.C. and Litchfield, D.W. (2000) The regulatory beta subunit of protein kinase CK2 mediates formation of tetrameric CK2 complexes. J. Biol. Chem. 275: 5003–5010.

Grasselli, E., Tomati, V., Bernasconi, M. V., Nicolini, C. & Vergani, L. (2004). C-terminal region of protein kinase CK2 alpha: How the structure can affect function and stability of the catalytic subunit. *Journal of Cellular Bbiochemistry*, Vol. 92, No. 2, (May 2004), pp. 270-284, ISSN 0730-2312

Guerra, B. (2006). Protein kinase CK2 subunits are positive regulators of AKT kinase. *International Journal of Oncology*, Vol. 28, No. 3, (March 2006), pp. 685-693, ISSN 1019-6439

Guerra B. & Issinger O.-G. (2008) Protein kinase CK2 in human diseases. *Current Medicinal Chemistry*, Vol. 15, No. 19, pp. 1870-1886, ISSN 0929-8673.

Guerra B., Issinger O.-G., Wang J.Y. (2003) Modulation of human checkpoint kinase Chk1 by the regulatory beta-subunit of protein kinase CK2. *Oncogene*, Vol. 22, No. 32, (August 7), pp. 4933-4942, ISSN 0950-9232.

Guerra, B., Boldyreff, B., Sarno, S., Cesaro, L., Issinger, O. G. & Pinna, L. A. (1999). CK2: a protein kinase in need of control. *Pharmacology & Therapeutics*, Vol. 82, No. 2-3, (May-June 1999), pp. 303-313, ISSN 0163-7258.

Gyenis, L. & Litchfield, D.W. (2008) The emerging CK2 interactome: insights into the regulation and functions of CK2. *Molecular and Cellular Biochemistry*, Vol. 316, No. 1-2, (September 2008), pp. 5-14, ISSN 0300-8177

Hanks, S. K. & Hunter, T. (1995). Protein kinases 6. The eukaryotic protein kinase superfamily: kinase (catalytic) domain structure and classification. *The FASEB Journal*, Vol. 9, No. 8, (May 1995), pp. 576-596, ISSN 0892-6638.

Hanks, S.K., Quinn, A.M., Hunter, T. (1988) The protein kinase family: conserved features and deduced phylogeny of the catalytic domains. *Science*, Vol. 241, No. 4861 , (July 1), pp. 42-52, ISSN 0036-8075.

Hinrichs, M. V., Jedlicki, A., Tellez, R., Pongor, S., Gatica, M., Allende, C. C. & Allende, J. E. (1993). Activity of recombinant alpha and beta subunits of casein kinase II from Xenopus laevis. *Biochemistry*, Vol. 32, No. 28, (July 1993), pp. 7310-7316, ISSN 0006-2960

Hinrichs, M. V., Gatica, M., Allende, C. C. & Allende, J. E. (1995). Site-directed mutants of the beta subunit of protein kinase CK2 demonstrate the important role of Pro-58. *FEBS Letters*, Vol. 368, No. 2, (July 1995), pp. 211-214, ISSN 0014-5793

Hu, E. & Rubin, C. S. (1990). Expression of wild-type and mutated forms of the catalytic (alpha) subunit of Caenorhabditis elegans casein kinase II in Escherichia coli. *The Jornal of Biological Chemistry,*, Vol. 265, No. 33, (November 1990), pp. 20609-20615, ISSN 0021-9258.

Issinger O.-G. (1993) Casein kinases: pleiotropic mediators of cellular regulation. *Pharmacology & Therapeutics*, Vol. 59, No. 1, (January 1993), pp. 1-30, ISSN 0163-7258.

Jacob, G., Neckelman, G., Jimenez, M., Allende, C. C. & Allende, J. E. (2000). Involvement of asparagine 118 in the nucleotide specificity of the catalytic subunit of protein kinase CK2. *FEBS Letters*, Vol. 466, No. 2-3, (January 2000), pp. 363-366, ISSN 0014-5793.

Jakobi, R. & Traugh, J. A. (1992). Characterization of the phosphotransferase domain of casein kinase II by site-directed mutagenesis and expression in Escherichia coli. *The Journal of Biological Chemistry*, Vol. 267, No. 33, (November 1992), pp. 23894-23902, ISSN 0021-9258.

Jakobi, R., Lin, W. J. & Traugh, J. A. (1994). Modes of regulation of casein kinase II. *Cellular & Molecular Biology Research*, Vol. 40, No. 5-6, pp. 421-429, ISSN 0968-8773.

Jakobi, R. & Traugh, J. A. (1995). Site-directed mutagenesis and structure/function studies of casein kinase II correlate stimulation of activity by the beta subunit with changes in conformation and ATP/GTP utilization. *European Journal of Biochemistry*, Vol. 230, No. 3, (June 1995), pp. 1111-1117, ISSN 0014-2956

Janeczko, M., Masłyk, M., Szyszka, R., Baier, A. (2011) Interactions between subunits of protein kinase CK2 and their protein substrates influences its sensitivity to specific inhibitors. *Molecular & Cellular Biochemistry*, Vol. 356, No. 1-2, (October 2011), pp. 121-126, ISSN 0300-8177.

Jauch, E., Melzig, J., Brkulj, M., Raabe, T. (2002) In vivo functional analysis of Drosophila protein kinase casein kinase 2 (CK2) beta-subunit. *Gene*, Vol. 298, No.1, (September 18), pp. 29-39, ISSN 0378-1119.

Jedlicki, A., Allende, C. C. & Allende, J. E. (2008). CK2alpha/CK1alpha chimeras are sensitive to regulation by the CK2beta subunit. *Molecular & Cellular Biochemistry*, Vol. 316, No. 1-2, (September 2008), pp. 25-35, ISSN 0300-8177

Jensen, B.C., Kifer, C.T., Brekken, D.L., Randall A.C., Wang, Q., Drees, B.L. & Parsons M. (2007) Characterization of protein kinase CK2 from *Trypanosoma brucei. Molecular & Biochemical Parasitology*, Vol. 151, No. 1, (January 2007), pp. 28-40, ISSN 0166-6851.

Johnson S.N. (2009) Protein kinase inhibitors: contributions from structure to clinical compounds. *Quarterly Reviews of Biophysics*, Vol. 42, No. 1, (February 2009), pp. 1-40, ISSN 0033-5835.

King, R.W., Glotzer, M., Kirschner, M.W. (1996) Mutagenic analysis of the destruction signal of mitotic cyclins and structural characterization of ubiquitinated intermediates. *Molecular Biology of the Cell*, Vol 7, No. 9, (September 1996), pp. 1343-1357, ISSN 1939-4586.

Knighton, D.R., Zheng, J.H., Ten Eyk, L.F., Ashford, V.A., Xuong, N.H., Taylor, S.S., Sowadski, J.M. (1991) Crystal structure of the catalytic subunit of cyclic adenosine monophosphate-dependent protein kinase. *Science*, Vol. 253, No. 5018, (July 1991), pp. 407-414, ISSN 0036-8075.

Kolaiti R.-M., Baier A., Szyszka R., Kouyanou-Koutsoukou S. (2011) Isolation of a CK2α subunit and the holoenzyme from the mussel *Mytilus galloprovincialis* and construction of the CK2α and CK2β cDNAs. *Marine Biotechnology (New York)*, Vol. 13, No. 3, (June 2011), pp. 505-516, ISSN 1436-2228.

Kopp, J., Schwede, T. (2004) Automated protein structure homology modeling: a progress report. *Pharmacogenomics*, Vol. 5, No. 4, (June 2004), pp. 405-416, ISSN 1462-2416.

Kouyanou-Koutsoukou S., Baier A., Kolaitis R.-M., Maniatopoulou E., Thanopoulou K., Szyszka R. (2011a) Cloning and purification of protein kinase CK2 recombinant alpha and beta subunits from the Mediterranean fly *Ceratitis capitata*. *Molecular and Cellular Biochemistry*, Vol. 356, No. 1-2, (October 2011), pp. 261-267, ISSN 0300-8177.

Kouyanou-Koutsoukou S., Kalpaxis D.L., Pytharopoulou S., Kolaitis R.-M., Baier A. and Szyszka R. (2011b) Translational control of gene expression in the mussel *Mytilus galloprovincials*: The impact of cellular stress on protein synthesis, the ribosomal stalk and the protein kinase CK2 activity. In: *Mussels: Anatomy, Habitat and Environmental Impact*. (Ed. Lauren E. McGevin), Nova Publisher, pp. 97-128, ISBN 9781617617638 1617617636.

Krehan, A., Lorenz, P., Plana-Coll, M. & Pyerin, W. (1996). Interaction sites between catalytic and regulatory subunits in human protein kinase CK2 holoenzymes as indicated by chemical cross-linking and immunological investigations. *Biochemistry*, Vol. 35, No. 15, (April 1996), pp. 4966-4975, ISSN 0006-2960

Kristensen, L.P., Larsen, M.R., Højrup, P., Issinger, O.-G., Guerra, B. (2004) Phosphorylation of the regulatory β-subunit of protein kinase CK2 by checkpoint kinase Chk1: identification of the *in vitro* CK2β phosphorylation site. *FEBS Letters*, Vol.569, No. 1-3, (July 2004), pp. 217-223, ISSN 0014-5793.

Krupa, A., Abhinandan, K.R., Srinivasan, N. (2004) KinG: a database of protein kinases in genomes. *Nucleic Acids Research*, Vol. 32 (Database Issue), (January 2004), pp. D153-D155, ISSN 0305-1048.

Kubiński, K., Domańska, K., Sajnaga, E., Mazur, E., Zieliński, R. & Szyszka, R. (2007) Yeast holoenzyme of protein kinase CK2 requires both β and β' regulatory subunits for its activity. *Molecular & Cellular Biochemistry*, Vol. 295, No. 1-2, (January 2007), pp. 229-235, ISSN 0300-8177.

Kubler, D., Pyerin, W., Burow, E. & Kinzel, V. (1983). Substrate-effected release of surface-located protein kinase from intact cells. *Proceedings of the National Academy of Sciences of the United States of America*, Vol. 80, No. 13, (July 1983), pp. 4021-4025, ISSN 0027-8424.

Kusk, M., Bendixen, C., Duno, M., Westergaard, O. & Thomsen, B. (1995). Genetic dissection of intersubunit contacts within human protein kinase CK2. *Journal of Molecular Biology*, Vol. 253, No. 5, (November 1995), pp. 703-711, ISSN 0022-2836.

Lebrin, F., Chambaz, E. M. & Bianchini, L. (2001). A role for protein kinase CK2 in cell proliferation: evidence using a kinase-inactive mutant of CK2 catalytic subunit alpha. *Oncogene*, Vol. 20, No. 16, (April 2001), pp. 2010-2022, ISSN 0950-9232.

Leroy, D., Alghisi, G.C., Roberts, E., Filhol-Cochet, O., Gasser, S.M. (1999) Mutations in the C-terminal domain of topoisomerase II affect meiotic function and interaction with the casein kinase 2 beta subunit. *Molecular & Cellular Biochemistry*, Vol. 191, No. 1-2, (January 1999), pp. 85-95, ISSN 0300-8177.

Leroy, D., Heriche, J.K., Filhol, O., Chambaz, E.M. Cochet, C. (1997) Binding of polyamines to an autonomous domain of the regulatory subunit of protein kinase CK2 induces a conformational change in the holoenzyme. A proposed role for the kinase stimulation. *The Journal of Biological Chemistry*, Vol. 272, No. 33, (August 1997), pp. 20820-20827, ISSN 0021-9258.

Li, D., Dobrowolska, G., Aicher, L. D., Chen, M., Wright, J. H., Drueckes, P., Dunphy, E. L., Munar, E. S. & Krebs, E. G. (1999). Expression of the casein kinase 2 subunits in Chinese hamster ovary and 3T3 L1 cells provides information on the role of the enzyme in cell proliferation and the cell cycle. *The Journal of Biological Chemistry*, Vol. 274, No. 46, (November 1999), pp. 32988-32996, ISSN 0021-9258

Li, D., Meier, U.T., Dobrowolska, G. & Krebs E.G. (1997) Specific interaction between casein kinase 2 and the nucleolar protein Nopp140. *The Journal of Biological Chemistry*, Vol. 272, No. 6, (February 1997), pp. 3773-3779, ISSN 0021-9258

Lieberman, S.L. & Ruderman, J.V. (2004) CK2 beta, which inhibits Mos function, binds to a discrete domain in the N-terminus of Mos. *Developmental Biology*, Vol. 268, No. 2, (April 2004), pp. 271-279, ISSN 0012-1606.

Liolli G., 2010. Structural dissection of cyclin dependent kinases regulation and protein recognition properties. *Cell Cycle*, Vol. 9, No. 8, (April 2010), pp. 1551-1561, ISSN 1538-4101.

Litchfield, D. W. (2003). Protein kinase CK2: structure, regulation and role in cellular decisions of life and death. *The Biochemical Journal*, Vol. 369, Pt 1, (January 2003), pp. 1-15, ISSN 0264-6021

Litchfield, D.W., Bosc, D.G., Slonimski, E. (1995) The protein kinase from mitotic human cells that phosphorylates Ser-209 on the casein kinase II beta-subunit is p34cdc2. *Biochemica et Biophysica Acta*, Vol. 1269, No. 1, (October 1995), pp. 69-78, ISSN 0006-3002.

Litchfield, D.W., Lozeman, F.J., Piening, C., Sommercorn, J., Takio, K., Walsh, K.A. & Krebs E.G. (1990) Subunit structure of casein kinase II from bovine testis: demonstration that the α and α´ subunits are distinct polypeptides. *The Journal of Biological Chemistry*, Vol. 265, No. 13, (May 1990), pp. 7638-7644, ISSN 0021-9258.

Luscher, B. & Litchfield, D. W. (1994). Biosynthesis of casein kinase II in lymphoid cell lines. *European Journal of Biochemistry*, Vol. 220, No. 2, (March 1994), pp. 521-526, ISSN 0014-2956.

Manning G., Plowman G.D., Hunter T. & Sudarsanam S. (2002a) Evolution of protein kinase signaling from yeast to man. *Trends in Biochemical Sciences*, Vol. 27, No. 10, (October 2002), pp. 514-520, ISSN 0968-0004

Manning, G., Whyte, D.B., Martinez, R., Hunter, T. & Sudarsanam S. (2002b) The protein kinase complement of the human genome. *Science*, Vol. 298, No. 5600, (December 2002), pp. 1912-1934, ISSN 0036-8075.

Maridor G., Park W., Krek W. & Nigg E.A. (1991) Casein kinase II. cDNA sequences, developmental expression and tissue distribution of mRNAs for a, a´ and b subunits of the chicken enzyme. *The Journal of Biological Chemistry*, Vol. 266, No. 4, (February 1991), pp. 2362-2368, ISSN 0021-9258.

Marin, O., Meggio, F. & Pinna, L.A. (1999) Structural features underlying the unusual mode of calmodulin phosphorylation by protein kinase CK2: A study with synthetic calmodulin fragments. *Biochemical and Biophysical Research Communications*, Vol. 256, No. 2, (March 1999), pp. 442–446, ISSN 0006-291X.

Miranda-Saavedra, D. & Barton, G.J. (2007) Classification and functional annotation of eukaryotic protein kinases. *Proteins*, Vol. 68, No. 4, (September 2007), pp. 893-914, ISSN 0887-3585.

Marin, O., Meggio, F., Draetta, G. & Pinna, L. A. (1992). The consensus sequences for cdc2 kinase and for casein kinase-2 are mutually incompatible. A study with peptides derived from the beta-subunit of casein kinase-2. *FEBS Letters*, Vol. 301, No. 1, (April 1992), pp. 111-114, ISSN 0014-5793.

Marin, O., Meggio, F., Boldyreff, B., Issinger, O. G. & Pinna, L. A. (1995). Dissection of the dual function of the beta-subunit of protein kinase CK2 ('casein kinase-2'): a synthetic peptide reproducing the carboxyl-terminal domain mimicks the positive but not the negative effects of the whole protein. *FEBS Letters*, Vol. 363, No. 1-2, (April 1995), pp. 111-114, ISSN 0014-5793.

Marin, O., Meggio, F., Sarno, S. & Pinna, L. A. (1997). Physical dissection of the structural elements responsible for regulatory properties and intersubunit interactions of protein kinase CK2 beta-subunit. *Biochemistry*, Vol. 36, No. 23, (June 1997), pp. 7192-7198, ISSN 0006-2960.

Meggio, F., Boldyreff, B., Marin, O., Marchiori, F., Perich, J. W., Issinger, O. G. & Pinna, L. A. (1992). The effect of polylysine on casein-kinase-2 activity is influenced by both the structure of the protein/peptide substrates and the subunit composition of the enzyme. *European Journal of Biochemistry*, Vol. 205, No. 3, (May 1992), pp. 939-945, ISSN 0014-2956.

Meggio, F., Boldyreff, B., Issinger, O. G. & Pinna, L. A. (1993). The autophosphorylation and p34cdc2 phosphorylation sites of casein kinase-2 beta-subunit are not essential for reconstituting the fully-active heterotetrameric holoenzyme. *Biochimica et Biophysica Acta*, Vol. 1164, No. 2, (July 1993), pp. 223-225, ISSN 0006-3002.

Meggio, F., Boldyreff, B., Issinger, O. G. & Pinna, L. A. (1994). Casein kinase 2 down-regulation and activation by polybasic peptides are mediated by acidic residues in the 55-64 region of the beta-subunit. A study with calmodulin as phosphorylatable substrate. *Biochemistry*, Vol. 33, No. 14, (April 1994), pp. 4336-4342, ISSN 0006-2960.

Meggio, F. & Pinna, L. A. (2003). One-thousand-and-one substrates of protein kinase CK2? *The FASEB journal : official publication of the Federation of American Societies for Experimental Biology*, Vol. 17, No. 3, (March 2003), pp. 349-368, ISSN 1530-6860

Messenger, M. M., Saulnier, R. B., Gilchrist, A. D., Diamond, P., Gorbsky, G. J. & Litchfield, D. W. (2002). Interactions between protein kinase CK2 and Pin1. Evidence for phosphorylation-dependent interactions. *The Journal of Biological Chemistry*, Vol. 277, No. 25, (June 2002), pp. 23054-23064, ISSN 0021-9258

Miller, S. J., Lou, D. Y., Seldin, D. C., Lane, W. S. & Neel, B. G. (2002). Direct identification of PTEN phosphorylation sites. *FEBS letters*, Vol. 528, No. 1-3, (September 2002), pp. 145-153, ISSN 0014-5793

Miravet, S., Piedra, J., Miro, F., Itarte, E., Garcia de Herreros, A. & Dunach, M. (2002). The transcriptional factor Tcf-4 contains different binding sites for beta-catenin and plakoglobin. *The Journal of Biological Chemistry*, Vol. 277, No. 3, (January 2002), pp. 1884-1891, ISSN 0021-9258

Nash, P., Tang, X., Orlicky, S., Chen, Q., Gertler, F. B., Mendenhall, M. D., Sicheri, F., Pawson, T. & Tyers, M. (2001). Multisite phosphorylation of a CDK inhibitor sets a threshold for the onset of DNA replication. *Nature*, Vol. 414, No. 6863, (November 2001), pp. 514-521, ISSN 0028-0836.

Niefind, K., Guerra, B., Ermakowa, I. & Issinger, O.-G. (2000). Crystallization and preliminary characterization of crystals of human protein kinase CK2. *Acta Crystallographica. Section D, Biological Crystallography*, Vol. 56, No. Pt 12, (December 2000), pp. 1680-1684, ISSN 0907-4449.

Niefind, K., Guerra, B., Ermakowa, I. & Issinger, O.-G. (2001). Crystal structure of human protein kinase CK2: insights into basic properties of the CK2 holoenzyme. *The EMBO journal*, Vol. 20, No. 19, (October 2001), pp. 5320-5331, ISSN 0261-4189.

Niefind, K. & Issinger, O. G. (2010). Conformational plasticity of the catalytic subunit of protein kinase CK2 and its consequences for regulation and drug design. *Biochimica et Biophysica Acta*, Vol. 1804, No. 3, (March 2010), pp. 484-492, ISSN 0006-3002.

Niefind K., Raaf J. & Issinger O.-G. (2009) Protein kinase CK2 in health and disease: Protein kinase CK2: from structures to insights. *Cellular and Molecular Life Sciences*, Vol. 66, No. 11-12, (June 2009), pp. 1800-1816, ISSN 1420-682X.

Olsen, B. B., Guerra, B., Niefind, K. & Issinger, O. G. (2010). Structural basis of the constitutive activity of protein kinase CK2. *Methods in Enzymology*, Vol. 484, No. pp. 515-529, ISSN 1557-7988.

Olsen, B. B. & Guerra, B. (2008) Ability of CK2β to selectively regulate cellular protein kinases. *Molecular & Cellular Biochemistry*, Vol. 316, No. 1-2, (September 2008), pp. 115-126, ISSN 0300-8177.

Olsten, M.E., Litchfield D.W., (2004) Order or chaos? An evaluation of the regulation of protein kinase CK2. *Biochemistry & Cell Biology*, Vol. 82, No. 6, (December 2004), pp. 681-693, ISSN 0829-8211.

Olsten, M.E., Weber J.E., Litchfield D.W. (2005) CK2 interacting proteins: emerging paradigms for CK2 regulation? *Molecular & Cellular Biochemistry*, Vol. 274, No. 1-2, (June 2005), pp. 115-124, ISSN 0300-8177.

Pagano, M.O., Bain, J., Kazimierczuk, Z., Sarno, S., Ruzzene, M., Di Maria, G., Elliott, M., Orzeszko, A., Cozza, G., Meggio, F. & Pinna, L.A. (2008) The selectivity of inhibitors of protein kinase CK2: an update. *Biochemical Journal*, Vol. 415, No. 3, (November 2008), pp. 353-365, ISSN 0264-6021.

Pandit, S.B., Balaji, S., Srinivasan,N. (2004) Structural and functional characterization of gene products encoded in the human genome by homology detection. *IUBMB Life*, Vol. 56, No. 6, (June 2004), pp. 317-331, ISSN 1521-6543.

Pearson, M.A. & Fabbro, D. (2004) Targetting protein kinases in cancer therapy: a success? *Expert Review of Anticancer Therapy*, Vol. 4, No.6., (December 2004), pp. 1113-1124, ISSN 1473-7140.

Pinna, L. A. (2002). Protein kinase CK2: a challenge to canons. *Journal of Cell Science*, Vol. 115, Pt. 20, (October 2002), pp. 3873-3878, ISSN 0021-9533

Pinna, L. A. & Meggio, F. (1997). Protein kinase CK2 ("casein kinase-2") and its implication in cell division and proliferation. *Progress in Cell Cycle Research*, Vol. 3, No. pp. 77-97, ISSN 1087-2957

Plowman, G.D., Sudarsanam, S., Bingham, J., Whyte, D., Hunter, T., (1999) The protein kinases of *Caenorhabditis elegans*, a model for signal transduction in multicellular organisms. *Proceedings of the National Academy of Sciences of the United States of America*, Vol. 96, No. 24, (November 1999), pp. 13603-13610, ISSN 0027-8424.

Ponce, D., Maturana, J. L., Cabello, P., Yefi, R., Niechi, I., Silva, E., Armisen, R., Galindo, M., Antonelli, M. & Tapia, J. C. (2011). Phosphorylation of AKT/PKB by CK2 is necessary for the AKT-dependent up-regulation of beta-catenin transcriptional activity. *Journal of Cellular Physiology*, Vol.226, No. 7, (July 2010), pp. 1953-1959, ISSN 1097-4652

Prudent, R., Sautel, C.F., Cochet, C. (2010) Structure-based discovery of small molecules targeting different surfaces of protein-kinase CK2. *Biochimica et Biophysica Acta*, Vol. 1804, No. 3, (March 2010), pp. 493-498, ISSN 0006-3002.

Raaf, J., Bischoff, N., Klopffleisch, K., Brunstein, E., Olsen, B. B., Vilk, G., Litchfield, D. W., Issinger, O.-G. & Niefind, K. (2011) Interaction between CK2alpha and CK2beta, the subunits of protein kinase CK2: thermodynamic contributions of key residues on the CK2alpha surface. *Biochemistry*, Vol. 50, No. 4, (February 2011), pp. 512-522, ISSN 1520-4995.

Raaf, J., Brunstein, E., Issinger, O.-G., Niefind, K. (2008) The interaction of CK2alpha and CK2beta, the subunits of protein kinase CK2, requires CK2beta in a preformed conformation and is enthalpically driven. *Protein Science*, Vol. 17, No. 12, (December 2008), pp. 2180-2186, ISSN 0961-8368.

Riera, M., Peracchia, G., de Nadal, E., Ariño, J., Pagès, M. (2001) Maize protein kinase **CK2**: regulation and functionality of three beta regulatory subunits. *The Plant Journal*, Vol. 25, No. 4, (February 2001), pp. 365-374, ISSN 0960-7412.

Rodriguez, F. A., Contreras, C., Bolanos-Garcia, V. & Allende, J. E. (2008). Protein kinase CK2 as an ectokinase: the role of the regulatory CK2beta subunit. *Proceedings of the National Academy of Sciences of the United States of America*, Vol. 105, No. 15, (April 2008), pp. 5693-5698, ISSN 1091-6490

Roussou, I, Draetta, G. (1994) The *Schizosaccharomyces pombe* casein kinase II alpha and beta subunits: evolutionary conservation and positive role of the beta subunit. *Molecular and Cellular Biology*, Vol. 14, No. 1, (January 1994), pp. 576-586, ISSN 0270-7306.

Sajnaga, E., Kubiński, K. & Szyszka, R. (2008). Catalytic activity of mutants of yeast protein kinase CK2alpha. *Acta Biochimica Polonica*, Vol. 55, No. 4, (November 2008), pp. 767-776, ISSN 0001-527X.

Salamon, J. A., Acuna, R. & Dawe, A. L. (2010). Phosphorylation of phosducin-like protein BDM-1 by protein kinase 2 (CK2) is required for virulence and G beta subunit stability in the fungal plant pathogen Cryphonectria parasitica. *Molecular Microbiology*, Vol. 76, No. 4, (May 2010), pp. 848-860, ISSN 1365-2958.

Sarno, S., Boldyreff, B., Marin, O., Guerra, B., Meggio, F., Issinger, O.-G. & Pinna, L. A. (1995). Mapping the residues of protein kinase CK2 implicated in substrate recognition: mutagenesis of conserved basic residues in the alpha-subunit.

Biochemical and Biophysical Research Communications, Vol. 206, No. 1, (January 1995), pp. 171-179, ISSN 0006-291X

Sarno, S., Ghisellini, P., Cesaro, L., Battistutta, R. & Pinna, L.A. (2001). Generation of mutants of CK2alpha which are dependent on the beta-subunit for catalytic activity. *Molecular & Cellular Biochemistry*, Vol. 227, No. 1-2, (November 2001), pp. 13-19, ISSN 0300-8177.

Sarno, S., Ghisellini, P. & Pinna, L.A. (2002). Unique activation mechanism of protein kinase CK2. The N-terminal segment is essential for constitutive activity of the catalytic subunit but not of the holoenzyme. *The Journal of Biological Chemistry*, Vol. 277, No. 25, (June 2002), pp. 22509-22514, ISSN 0021-9258.

Sarno, S., Marin, O., Boschetti, M., Pagano, M.A., Meggio, F., Pinna, L.A. (2000) Cooperative modulation of protein kinase CK2 by separate domains of its regulatory beta-subunit. *Biochemistry*, Vol. 39, No.40, (October 10), pp. 12324-12329, ISSN 0006-2960.

Sarno, S., Ruzzene, M., Frascella, P., Pagano, M. A., Meggio, F., Zambon, A., Mazzorana, M., Di Maira, G., Lucchini, V. & Pinna, L. A. (2005). Development and exploitation of CK2 inhibitors. *Molecular & Cellular Biochemistry*, Vol. 274, No. 1-2, (June 2005), pp. 69-76, ISSN 0300-8177.

Sarno, S., Salvi, M., Battistutta, R., Zanotti, G. & Pinna, L.A. (2005). Features and potentials of ATP-site directed CK2 inhibitors. *Biochimica et Biophysica Acta*, Vol. 1754, No. 1-2, (December 2005), pp. 263-270, ISSN 0006-3002.

Sarno, S., Vaglio, P., Cesaro, L., Marin, O. & Pinna, L.A. (1999). A multifunctional network of basic residues confers unique properties to protein kinase CK2. *Molecular & Cellular Biochemistry*, Vol. 191, No. 1-2, (January 1999), pp. 13-19, ISSN 0300-8177.

Sarno, S., Vaglio, P., Marin, O., Meggio, F., Issinger, O.-G. & Pinna, L.A. (1997a). Basic residues in the 74-83 and 191-198 segments of protein kinase CK2 catalytic subunit are implicated in negative but not in positive regulation by the beta-subunit. *European Journal of Biochemistry*, Vol. 248, No. 2, (September 1997), pp. 290-295, ISSN 0014-2956.

Sarno, S., Vaglio, P., Marin, O., Issinger, O. G., Ruffato, K. & Pinna, L.A. (1997b). Mutational analysis of residues implicated in the interaction between protein kinase CK2 and peptide substrates. *Biochemistry*, Vol. 36, No. 39, (September 1997), pp. 11717-11724, ISSN 0006-2960.

Sarno, S., Vaglio, P., Meggio, F., Issinger, O. G. & Pinna, L. A. (1996). Protein kinase CK2 mutants defective in substrate recognition. Purification and kinetic analysis. *The Journal of Biological Chemistry*, Vol. 271, No. 18, (May 1996), pp. 10595-10601, ISSN 0021-9258.

Seldin, D. C., Landesman-Bollag, E., Farago, M., Currier, N., Lou, D. & Dominguez, I. (2005). CK2 as a positive regulator of Wnt signalling and tumourigenesis. *Molecular & Cellular Biochemistry*, Vol. 274, No. 1-2, (June 2005), pp. 63-67, ISSN 0300-8177.

Shi, X., Potvin, B., Huang, T., Hilgard, P., Spray, D.C., Suadicani, S.O., Wolkoff, A.W., Stanley, P. and Stockert, R.J. (2001) A novel casein kinase 2 alpha-subunit regulates membrane protein traffic in the human hepatoma cell line HuH-7. *The Journal of Biological Chemistry*, Vol. 276, No. 3, (January 2001), pp. 2075-2082, ISSN 0021-9258 276: 2075-2082.

Song, D. H., Dominguez, I., Mizuno, J., Kaut, M., Mohr, S. C. & Seldin, D. C. (2003). CK2 phosphorylation of the armadillo repeat region of beta-catenin potentiates Wnt signaling. *The Journal of Biological Chemistry*, Vol. 278, No. 26, (June 2003), pp. 24018-24025, ISSN 0021-9258.

Srinivasan, N., Antonelli, M., Jacob, G., Korn, I., Romero, F., Jedlicki, A., Dhanaraj, V., Sayed, M. F., Blundell, T. L., Allende, C. C. & Allende, J. E. (1999). Structural interpretation of site-directed mutagenesis and specificity of the catalytic subunit of protein kinase CK2 using comparative modelling. *Protein Engineering*, Vol. 12, No. 2, (February 1999), pp. 119-127, ISSN 0269-2139.

St-Denis, N.A., Derksen, D.R. & Litchfield, D.W. (2009) Evidence for regulation of mitotic progression through temporal phosphorylation and dephosphorylation of CK2alpha. *Molecular & Cellular Biology*, Vol. 29, No. 8, (April), pp. 2068-2081, ISSN 0270-7306.

Trembley J.H., Wang G., Unger G., Slaton J. & Ahmed K. (2009) CK2: a key player in cancer biology. *Cellular and Molecular Life Sciences*, Vol. 66, No 11-12, (June 2009), pp. 1858-1867, ISSN 1420-682X.

Tuazon, P. T. & Traugh, J. A. (1991). Casein kinase I and II--multipotential serine protein kinases: structure, function, and regulation. *Advances in Second Messenger and Phosphoprotein Research*, Vol. 23, No. (January 1991), pp. 123-164, ISSN 1040-7952.

Vaglio, P., Sarno, S., Marin, O., Meggio, F., Issinger, O. G. & Pinna, L. A. (1996). Mapping the residues of protein kinase CK2 alpha subunit responsible for responsiveness to polyanionic inhibitors. *FEBS letters*, Vol. 380, No. 1-2, (February 1996), pp. 25-28, ISSN 0014-5793.

Verma, R., McDonald, H., Yates, J. R., 3rd & Deshaies, R. J. (2001). Selective degradation of ubiquitinated Sic1 by purified 26S proteasome yields active S phase cyclin-Cdk. *Molecular Cell*, Vol. 8, No. 2, (August 2001), pp. 439-448, ISSN 1097-2765.

Vilk, G., Derksen, D. R. & Litchfield, D. W. (2001). Inducible expression of the regulatory protein kinase CK2beta subunit: incorporation into complexes with catalytic CK2 subunits and re-examination of the effects of CK2beta on cell proliferation. *Journal of Cellular Biochemistry*, Vol. 84, No. 1, pp. 84-99, ISSN 0730-2312.

Vilk G., Weber J.E., Turowiec J.P., Duncan J.S., Wu C., Derksen D.R., Zień P., Sarno S., Donella-Deana A., Lajoie G., Pinna L.A., Li S.S. and Litchfield D.W. (2008) Protein kinase CK2 catalyzes tyrosine phosphorylation in mammalian cells. *Cellular Signalling*, Vol. 20, No. 11, (November 2008), pp. 1942-1951, ISSN 0898-6568.

Wirkner, U., Voss, H., Lichter, P., Pyerin, W. (1994) Human protein kinase CK2 genes. *Cellular & Molecular Biology Research*, Vol. 40, No. 5-6, pp 489-99., ISSN 0968-8773.

Yde, C. W., Ermakova, I., Issinger, O. G. & Niefind, K. (2005). Inclining the purine base binding plane in protein kinase CK2 by exchanging the flanking side-chains generates a preference for ATP as a cosubstrate. *Journal of Molecular Biology*, Vol. 347, No. 2, (March 2005), pp. 399-414, ISSN 0022-2836

Zhang, C., Vilk, G., Canton, D. A. & Litchfield, D. W. (2002). Phosphorylation regulates the stability of the regulatory CK2beta subunit. *Oncogene*, Vol. 21, No. 23, (May 2002), pp. 3754-3764, ISSN 0950-9232

Zhang, N. & Zhong, R. (2010). Structural basis for decreased affinity of Emodin binding to Val66-mutated human CK2 alpha as determined by molecular dynamics. *Journal of Molecular Modeling*, Vol. 16, No. 4, (April 2010), pp. 771-780, ISSN 0948-5023

Zhang, Y. & Dong, C. (2007) Regulatory mechanisms of mitogen-activated kinase signaling. *Cellular & Molecular Life Sciences*, Vol. 64, No. 21, (November 2007), pp. 2771-2789, ISSN 1420-682X.

8

Directed Mutagenesis in Structure Activity Studies of Neurotransmitter Transporters

Jane E. Carland, Amelia R. Edington, Amanda J. Scopelliti,
Renae M. Ryan and Robert J. Vandenberg
Department of Pharmacology, The University of Sydney
Australia

1. Introduction

The delicate balance between excitation and inhibition within the central nervous system is critical to the maintenance of normal brain function. Players key to this balance are neurotransmitter transporters. Neurotransmitter transporters are drawn from two families of solute carriers (SLC), SLC1 and SLC6. The transporters for glutamate and small neutral amino acids belong to the SLC1 family, while transport of monoamines (5-hydroxytryptamine, dopamine, noradrenaline) and amino acid neurotransmitters (γ-aminobutyric acid, glycine) are mediated by members of the SLC6 family. These integral membrane proteins regulate the concentration of neurotransmitters, such as glutamate and glycine, within the synapse. They utilise pre-existing electrochemical gradients to drive the transport of neurotransmitters across neuronal and glial membranes, terminating neurotransmission and replenishing intracellular levels of neurotransmitter for future release. Neurotransmitter transporters are targeted by a number of substances, both therapeutic (antidepressants, anticonvulsant, antipsychotics, analgesics, anxiolytics) and addictive (cocaine, methampetamine). Their dysfunction is associated with multiple disorders, including epilepsy, ischaemic stroke, neuropathic pain and schizophrenia (Dohi et al., 2009; Sur & Kinney, 2004). Thus, structure activity studies of transporters are essential to provide new insights into their function and direct the design of novel, transporter-specific therapeutics.

Since the cloning of the GABA transporter, GAT1, in 1990 (Guastella et al., 1990), directed mutagenesis studies have underpinned our understanding of the secondary structure and function of neurotransmitter transporters. This work has subsequently been supported, and significantly advanced, by the high resolution crystal structures of prokaryotic homologues of the SLC1 and SLC6 families. The crystal structure of the SLC1 homologue from *Pyrococcus horikoshii* (Glt$_{Ph}$) was the first to be solved at 3.5 Å resolution (Fig 1A) (Yernool et al., 2004). This was followed in 2005 by the crystal structure of a homologue of the SLC6 family from *Aquifex aeolicus* (LeuT$_{Aa}$) at 1.65 Å resolution (Yamashita et al., 2005) (Fig 3A). These crystal structures have provided important insights into the interactions of transporters with substrates, ions, lipids and inhibitors, allowing the postulation of numerous functional mechanisms. However, the details provided by these high-resolution structures are insufficient to fully understand transport mechanisms. The knowledge obtained from crystal structures and 3D models can be used to direct mutagenesis work (utilising chimeric transporters and site-directed mutagenesis), electrophysiology and uptake studies to determine the molecular basis for transporter function. The techniques described in this chapter can be applied to other membrane proteins, such as G-protein coupled receptors and ligand-gated ion channels.

1.1 The SLC1 family of transporters

The SLC1 family of neurotransmitter transporters includes five human excitatory amino acid transporters (EAAT1 to EAAT5) (Slotboom et al., 1999) and two neutral amino acid transporters (ASCT1 and ASCT2) (Arriza et al., 1993; Kanai & Hediger, 2004). The EAATs exhibit 40-44% sequence identity with ASCTs and approximately 36% sequence identity with the related Na^+-coupled aspartate transporter, Glt$_{Ph}$ (Fig. 2). The high resolution crystal structure of Glt$_{Ph}$ reveals that SLC1 transporters exist as bowl-shaped trimers (Fig. 1A) (Yernool et al., 2004). Each protomer is comprised of eight transmembrane domains (TM1-8) and two re-entrant hairpin loops (HP1 and HP2). TM1, TM2, TM4, and TM5 mediate intersubunit contacts and support the "transport" domain, which is composed of TM3, TM6, TM7, TM8, HP1 and HP2 (Fig. 1B). This transport domain mediates substrate and ion translocation, and each protomer has an independent translocation pathway.

The EAATs are critical components of excitatory synapses, where they mediate the high affinity uptake of the dominant excitatory neurotransmitter, glutamate, as well as L- and D-aspartate. Both EAAT1 and EAAT2 are expressed in glia. Of these two subtypes, EAAT2 is the more widely distributed and is the major regulator of glutamate concentrations in the central nervous system. EAAT3 is expressed on neuronal membranes throughout the brain, while EAAT4 is selectively expressed on cerebellar Purkinje cells. EAAT5 is expressed on retinal neurons. Glutamate uptake is coupled to the co-transport of three Na^+ ions and one H^+ and the counter-transport of one K^+ ion, rendering them electrogenic (Zerangue & Kavanaugh, 1996). In addition, glutamate transport is associated with a thermodynamically uncoupled Cl^- conductance (Fig. 1C) (Fairman et al., 1995, Wadiche et al., 1995).

1.2 The SLC6 family of transporters

Members of the SLC6 transporter family are responsible for the transport of monoamine (dopamine, serotonin/5-hydroxytryptamine, noradrenaline) and amino acid (GABA and glycine) neurotransmitters across cell membranes. Two glycine transporters (GLYT1 and GLYT2) have been cloned, and five GLYT1 splice variants (GLYT1a to GLYT1e) and three GlyT2 splice variants (GLYT2a to GLYT2c) have been identified. The crystal structure of the prokaryotic transporter LeuT$_{Aa}$ serves as a useful template for unravelling the functional implications of transporter structures (FIG. 3). Members of the SLC6 family are traditionally thought to exist as monomers (Horiuchi et al., 2001; Lopez-Corcuera et al., 1993), although more recent work suggests that they may form dimers in vivo (Bartholomaus et al., 2008). Each subunit is formed by twelve transmembrane domains (TM1-12), with amino- and carboxy-termini located on the intracellular side of the membrane. Each subunit exhibits a two-fold axis of symmetry, with TM1 to TM5 corresponding to TM6 to TM10 with an inverted topology repeat. TM1 to TM10 form the core of the transporter, with TM1 and TM6 exhibiting the highest degree of sequence homology. They run antiparallel and exist with a central unwound section. The area surrounding this central unwound section is critical for substrate and ion binding.

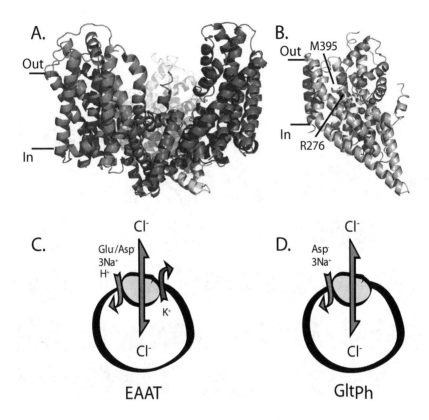

Fig. 1. The crystal structure of the prokaryotic transporter Glt$_{Ph}$ and the stoichiometry of transport by the excitatory amino acid transporters (EAAT1- EAAT5) and Glt$_{Ph}$. A. Glt$_{Ph}$ is a bowl shaped trimer viewed in the plane of the membrane. Individual protomers are coloured red, green and blue. B. A single protomer of the Glt$_{Ph}$ trimer (PDB 2NWX). The C-terminal domain is shown in colour; HP1 (yellow), TM7 (orange), HP2 (red) and TM8 (magenta). Bound aspartate is shown in stick representation and two Na$^+$ ions are shown as blue spheres. R276 (HP1) and M395 (TM8), which are discussed in section 3.5, are also shown in stick representation. Structures were viewed and rendered in PyMOL (http://www.pymol.org) (Schrodinger, 2010). C. Glutamate or aspartate transport via the EAATs is coupled to three Na$^+$ ions and one H$^+$, followed by the counter-transport of one K$^+$ ion. Binding of Na$^+$ and substrate to the EAATs activates a thermodynamically uncoupled Cl$^-$ conductance (*pink arrow*). D. Aspartate transport via Glt$_{Ph}$ is coupled to the co-transport of three Na$^+$ ions, but is not coupled to the movement of either H$^+$ or K$^+$ ions. Na$^+$ and aspartate binding to Glt$_{Ph}$ also activates an uncoupled Cl$^-$ conductance (*pink arrow*).

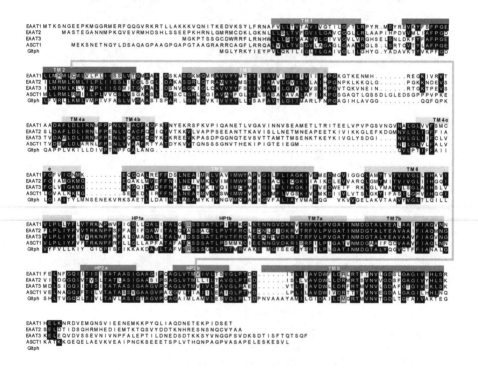

Fig. 2. Amino acid sequence alignment of SLC1 family members. Sequences for EAAT1-3, ASCT1 and Glt_{Ph} are shown. Transmembrane domains are indicated using the colour scheme as for the structure of Glt_{Ph} in Fig. 1. Homologous regions are highlighted in black. Residues highlighted in red boxes with yellow background are discussed in the text. The blue line (connecting Q93 to V452 in EAAT1) indicates that cysteine mutants of these two residues can be cross-linked (see section 3.4 for details).

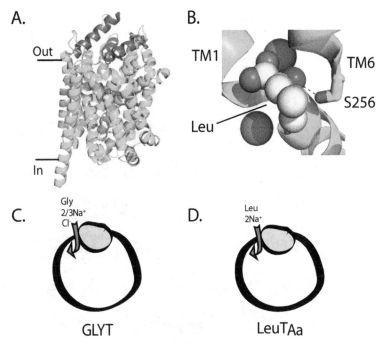

Fig. 3. The crystal structure of the prokaryotic transporter LeuT$_{Aa}$ and the stoichiometry of transport by the GLYTs and LeuT$_{Aa}$. A. The structure of LeuT$_{Aa}$ viewed in the plane of the membrane (PDB 2A65). Bound leucine is shown in A space-filling representation, and the 2 Na$^+$ ions are shown as purple spheres. Extracellular loop 2 (EL2, grey) and extracellular loop 4 (EL4, blue) are highlighted (see section 2.2 for details). EL4 residues R531 and K532 (orange stick representation) and I545 (red stick representation) are also highlighted (see section 3.2 for discussion). B. Bound leucine interacts with transmembrane domains TM1 and TM6. The hydrogen bond between the nitrogen atom of leucine and the side chain of serine 265 is represented by a dashed line (see section 3.2 for details). Na$^+$ ions are shown as purple spheres. C. Glycine transport via the GLYTs is coupled to two (GLYT1) or three (GLYT2) Na$^+$ ions and one Cl$^-$ ion. D. Alanine transport via LeuT$_{Aa}$ is coupled to the co-transport of two Na$^+$ ions.

Mammalian members of the SLC6 family share 20-25% sequence identity with the prokaryotic transporter LeuT$_{Aa}$ (FIG. 4). GLYT1 and GLYT2 are structurally similar and exhibit 48% sequence homology (FIG. 4), but they display significant functional differences. GLYT1 is predominantly expressed in glia at excitatory glutamatergic synapses, where they are responsible for regulating glycine, which acts as a co-agonist at NMDA receptors. Of the five GLYT1 isomers identified, GLYT1b and GLYT1c are nervous system-specific. In contrast, GLYT2 is typically localized with glycine receptors at inhibitory glycinergic synapses (spinal cord). The translocation of glycine by both transporters is coupled to the co-transport of Na$^+$ and Cl$^-$ (FIG. 3), but the stoichiometry of ion-flux coupling by GLYT1 and GLYT2 differs. The transport of one glycine molecule is coupled to the co-transport of two Na⁺ ions and one Cl⁻ ion for GLYT1 transporters, while the movement of three Na⁺ ions and one Cl⁻ ion is coupled to glycine transport for GLYT2 (Fig. 3C).

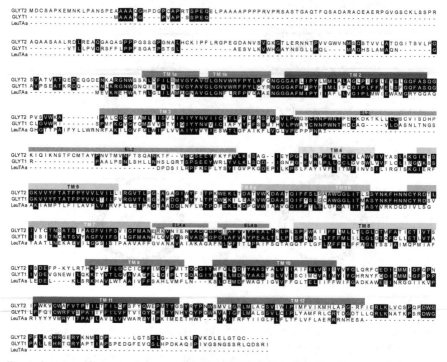

Fig. 4. Amino acid sequence alignment of GLYT1, GLYT2 and LeuT$_{Aa}$. Transmembrane domains are indicated using the colour scheme used for the structure of LeuT$_{Aa}$. Homologous regions are highlighted in black. Chimeric transporters were generated between GLYT1 and GLYT2 in which extracellular loops 2 and 4 (highlighted with yellow edges) were switched between the two transporters (see section 2.2 for details). Residues highlighted in red boxes with yellow background are discussed in the text.

2. The use of chimeras in studies of neurotransmitter transporters

Chimeras provide an excellent tool for the study of neurotransmitter transporters. Switching specific regions/structures between transporters can provide insights into transporter function and substrate selectivity and inform the design of directed mutagenesis studies. Further, chimeras between mammalian and bacterial transporters have the potential to assist with the crystallisation of transporters, thereby facilitating the determination of the structure of mammalian transporters. We employ a fusion PCR technique to create chimeric junctions at specific amino acid locations, allowing for the precise design of chimeras. We have produced numerous chimeras using this method, including chimeras of GLYT1 and GLYT2.

2.1 PCR fusion methodology

Conventional chimera construction methodology relies on restriction enzyme cloning, in which unique restriction enzyme sites are introduced into both the acceptor and the donor proteins. While this technique allows for the production of chimeras with specific/known junction points, the availability of unique restriction enzyme sites can impose limitations on the design of potential chimeras. In contrast, the PCR fusion technique (Shevchuk et al., 2004) creates chimeric junctions at any amino acid, without the need for restriction enzyme sites. In this method (Fig. 5), each segment of the final chimera is amplified in individual PCR reactions. The primers are designed to engineer complementary overlapping sequences onto the junction-forming ends of each product. The PCR primers possess typical properties (18-24 nucleotides in length and a melting temperature of ~64°C). The overlapping sequences correspond to the desired chimeric junction sites between subunits. The strands of the PCR products (duplex DNAs) are routinely separated and allowed to reanneal by cycles of heating and cooling. A partially duplex chimera can form as a result of annealing by the complementary regions of different fragments, one half of the chimeras will have free 3'-ends that can be elongated by DNA polymerase. The other half will have free 5'-ends that cannot be elongated. If more than two PCR products are involved, eventually a full-length chimera forms. The resulting duplex chimera is then amplified by PCR with oligonucleotide primers containing restriction enzyme sites at the 5' and 3' ends of the DNA of the chimera to facilitate subcloning into a suitable vector. For chimeras that require more than three fragments, it is often best to produce an initial chimera of three fragments and then incorporate additional fragments as required. It is important to obtain complete DNA sequences of any clones generated in this manner to confirm the junction sites and also to ensure that there have been no spurious sequence changes.

2.2 Identifying determinants of drug selectivity

Drugs that have selective effects *in vivo* are much sought after. Such compounds potentially have minimal side effects, making them attractive options as therapeutics. Studies suggest that the GLYTs may provide a novel therapeutic target for the development of drugs to treat neurological disorders and pain. In particular, GLYT1 is considered a potential target for the development of agents to treat schizophrenia (Sur et al., 2004). GLYT2 is a key target for studies to develop molecules to treat chronic pain (Aragon & Lopez-Corcuera, 2003). N-Arachidonylglycine (NAGly) is an endogenous derivative of arachidonic acid. NAGly has been shown to induce analgesia in rat models of neuropathic and inflammatory pain (Succar et al., 2007; Vuong et al., 2008). One of the mechanisms of action of NAGly is the inhibition of GLYT2 (IC_{30} = 3 µM), whilst it has no effect on GLYT1 (Wiles et al., 2006). Understanding the molecular basis of NAGly selectivity may aid in the development of novel analgesic compounds.

Extracellular loops EL2 and EL4 have been implicated in mediating inhibition of the GLYTs. Residues that contribute to the binding site of Zn^{2+}, a non-competitive inhibitor of GLYT1, have been identified in EL2 and EL4 (Ju et al., 2004). Zn^{2+} has been proposed to inhibit GLYT1 by binding to EL2 and EL4, restricting the movement of these loops and thus preventing glycine transport. LeuT$_{Aa}$, the bacterial homologue of the GLYTs, has been crystallized in the presence of clomipramine, a non-competitive inhibitor (Singh et al., 2007), and tryptophan, a competitive inhibitor (Singh et al., 2008). Clomipramine was shown to

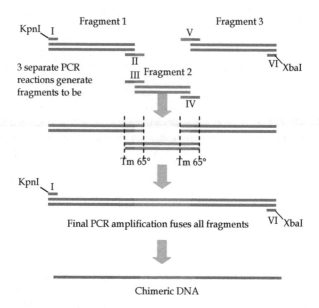

Fig. 5. A schematic summary of the construction of chimera GLYT1(EL2) using the PCR fusion methodology. The DNA sequence of the donor cDNA (GLYT1) is in blue and the acceptor cDNA (GLYT2) is in red. cDNAs of GLYT1 and GLYT2 in the presence of their corresponding primers (I and II, III and IV, V and VI) undergo PCR to generate three fragments with the appropriate homologous ends. The fragments are 'fused' together in another reaction, in which the single strands from the overlapping regions serve as internal primers (see text). The final PCR amplification reaction fuses ('zips') all the fragments in the presence of primers I and VI, which, in this case, include restriction site for the enzymes KpnI and XbaI. The final product is the chimera GLYT1(EL2) with two unique restriction sites engineered on either end to allow for insertion of the chimeric gene into the vector pOTV. The restriction sites may be altered for subcloning into different vectors.

stabilize LeuT$_{Aa}$ in an occluded state by interacting with a number of transmembrane domains and displacing the tip of EL4. In contrast, the presence of the competitive inhibitor tryptophan appeared to trap the transporter in an open-to-out conformation. Four separate tryptophan molecules were identified in this crystal structure, with one of them found to be interacting with EL2 and EL4. These observations suggest that these loops play key roles in the inhibition of GLYTs.

In order to ascertain the molecular basis for the inhibitory activity of NAGly on GLYT2, chimeras were generated between GLYT1 and GLYT2, in which EL2 and/or EL4 were switched (Fig. 6). A chimera is named according to its parental transporter, with the inserted loop in parentheses. For example, GLYT2(EL2) is predominantly GLYT2 with the EL2 of GLYT1. One of the important controls required when using a chimeric protein to understand structure-function relationships is the ability to maintain the functional properties of the chimera. For chimeras that are predominantly GLYT2, the EC$_{50}$ for glycine was very similar to that of the parental GLYT2 transporter, which indicates that these chimeras transport glycine similar to GLYT2. With the GLYT1-based chimeras, GLYT1(EL2) and GLYT1(EL2, EL4), the EC_{50} for glycine was increased by 6-8 fold compared to GLYT1. Thus, for the GLYT1-based chimeras, the transport of glycine differs slightly from the parent transporter; and, therefore, we are not able to be as confident that other functional changes are solely due to the region of interest. Nevertheless, the experiments using the GLYT2-based chimeras yielded valuable information concerning the domains responsible for NAGly sensitivity. GLYT2 is inhibited by NAGly, whereas GLYT2(EL2) and GLYT2(EL4) have reduced sensitivity (Edington et al., 2009). These observations suggest that EL2 and EL4 contribute to forming an inhibitory site on GLYT2.

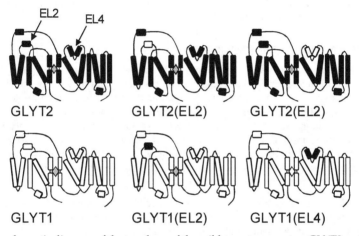

Fig. 6. A schematic diagram of the topology of the wild-type transporters GLYT2 and GLYT1 and their EL2 and EL4 chimeras. *Black* indicates the regions of the transporter from GLYT2 and *white* indicates the regions from GLYT1. Shown are GLYT2, GLYT2(EL2), GLYT2(EL4), GLYT1, GLYT1(EL2), and GLYT1(EL4).

3. Site-directed mutagenesis

Identifying domains that are responsible for conferring functional differences between transporters is only the first step in understanding the molecular processes that dictate neurotransmitter transporter function. Chimeric and directed mutagenesis studies are mutually beneficial and, when used in conjunction, can increase the efficiency and efficacy of experimentation. Information obtained in the study of chimeric transporters helps to focus site-directed mutagenesis studies to specific domains. The introduction of point mutations within these regions can then provide very useful information about the function of specific residues and the location of substrate and ion binding sites. Conventional mutagenesis studies utilise amino acid sequence alignments between multiple members of a transporter family to direct and design mutagenesis studies. However, this work often relies on the mutation knocking out a particular function and thereby assigning that function to the residue of interest. This approach can generate misleading conclusions because loss of function can be the result of a number of changes, many of which do not necessarily reflect a direct disruption of the interaction being investigated. To avoid this issue, we employ knowledge obtained from recent advances in high resolution transporter crystal structures and homology models to improve our accuracy in predicting residues of interest and successfully creating functional mutants. Crystal structures provide a powerful 3-dimensional tool that enables us to visualise each residue and their individual contacts with the substrate/ions/other residues, thus improving the selection process. In addition, homology models provide a computational prediction of the global effect of a mutation. Thus, they can be used to predict which mutations will be accommodated by individual transporters. We have undertaken numerous mutagenesis studies that have, among other things, identified the glycine substrate binding site in GLYT1 and GLYT2, helped to characterise selective drug-binding sites on the GLYTs and clarified the specificity of ion and substrate interactions with the EAATs and Glt$_{Ph}$.

3.1 Directed mutagenesis methodology

Site-directed mutagenesis is a powerful tool that makes select changes to the genetic code of a protein to alter the amino acid sequence. The resultant mutant protein exhibits subtle structural changes. The ability to selectively manipulate amino acids in a controlled manner provides a powerful tool that is exploited by researchers undertaking structure-function studies. It should be noted that the importance of loss of function mutants must be interpreted with caution. Disruption of transport functions following a mutation can result in loss of function. However, loss of function could also result from misfolding of a protein or altered expression levels. Therefore, mutants that lose function are only useful if the cause of loss of function can be accurately ascertained.

Site-directed mutagenesis involves the use of specifically designed primers (sense and antisense) that include a point mutation of interest in the centre of the sequence. Each primer is 18-24 nucleotides in length with a melting temperature of ~64°C. The primers incorporate the mutation into an intronless gene, using the cDNA as template in a PCR reaction, with the elongation by DNA polymerase. Following amplification, the endonuclease DpnI, which recognizes methylated and hemimethylated DNA, can be added to the amplification reaction to digest the parental cDNA. It is important to obtain DNA sequences to confirm the mutation. The mutant DNA is subcloned into the appropriate vector. In the studies we describe, the Oocyte Transcription Vector (pOTV) is used, as it enables efficient RNA production and facilitates high expression levels of protein in *Xenopus laevis* oocytes. To construct a double point mutation, one set of primers may be used when the two residues are next to each other or in close proximity. However, if the two mutations are further apart, DNA encoding one mutation is initially made and used as the template for the second mutation. Site-directed mutagenesis kits are sold by many companies (e.g., Stratagene, Promega, Clonetech). We routinely use the QUIKCHANGE(tm) Site-Directed Mutagenesis kit from Stratagene.

3.2 Mutagenesis to identify determinants of drug selectivity of glycine transporters

The use of chimeric GLYT transporters allowed the identification of EL2 and EL4 as being regions critical for determining the selective activity of NAGly on GLYT2 (Edington et al., 2009). Having identified the regions of interest, we used site-directed mutagenesis to identify the residues within these regions that form the selective drug-binding site. Our chimeric study (see Section 2.2) revealed that NAGly interacted with EL4, so the aim of the subsequent mutagenesis study was to identify the key residues in EL4 that play a role in determining the differential sensitivity of GLYT2 compared to GLYT1.

The EL4 of GLYT2 exhibits 60% sequence identity with the GLYT1 EL4. Point mutations were introduced at all positions within the GLYT2 EL4 that differed from GLYT1. In total, eleven mutations were produced. Each resulted in a functional mutant GLYT2 transporter that exhibited glycine transport unchanged from that of the wild-type transporter. Three mutations resulted in changes in sensitivity to NAGly. Mutation of arginine at position 531 to leucine (R531L) and lysine at position 532 to glycine (K532G) resulted in modest reductions in NAGly sensitivity (IC_{30} = 13 ± 2 μM and 9 ± 1 μM, respectively) compared to GLYT2 (IC_{30} = 3.4 ± 0.6 μM). In contrast, the mutation of isoleucine at position 545 to leucine (I545L) markedly reduced NAGly sensitivity (IC_{30}>30μM) (Edington et al., 2009).

This work revealed that three residues within EL4 are critical to the inhibitory activity of NAGly on GLYT2. Modelling studies place I545 in the middle of EL4, while R531 and K532 are located at the edge of EL4. It is surprising that a conservative mutation at position 545 (an isoleucine to a leucine) would have the most dramatic impact. Two likely possibilities follow. (1) NAGly may bind to I545, and the I545L mutation may distort the way that NAGly fits into the binding site. (2) The I545L mutation may alter the conformation of the two arms of EL4, which then impacts on the way that this domain interacts with other elements that may be crucial for NAGly binding. The carboxyl groups of NAGly may interact with the positively charged R531 and K532 residues. Further structural studies are required to fully characterize the specific interaction sites between the NAGly and GLYT2.

3.3 Substrate selectivity of the GLYTs

GLYT1 and GLYT2 can be differentiated by their substrate selectivity and inhibitor sensitivity. GLYT1 transports both glycine and the N-methyl derivative of glycine, sarcosine, while GLYT2 only transports glycine (Supplisson & Bergman, 1997). In addition, GLYT1 is selectively inhibited by N[3-(4-fluorophenyl)-3-(4'-phenylphenoxy) propylsarcosine (NFPS) (Aubrey & Vandenberg, 2001). The crystal structure of LeuT$_{Aa}$ provides a good working model for the study of GLYTs. As with the GLYTs, the substrate binding site of LeuT$_{Aa}$ is formed at the junction between TM1–5 and TM6–10. It is composed of amino acid residues from TM1 and TM6 (Yamashita et al., 2005). Both of these transmembrane domains contain an unwound segment, and many of the substrate contact sites are with the main chain atoms of these unwound segments. Sequence alignments between LeuT$_{Aa}$ and the GLYTs revealed that there are a number of identical residues and some key differences in the predicted substrate binding site. In particular, in the crystal structure of LeuT$_{Aa}$, the amino group of leucine is hydrogen-bonded to the hydroxyl group of the side chain of serine at position 256 (Fig 3B). This residue is located in TM6. We focused on the role of the corresponding residues in GLYT1b and GLYT2a to investigate if this residue is a determinant of GLYT substrate selectivity (Vandenberg et al., 2007).

Serine at position 256 in LeuT$_{Aa}$ corresponds to a glycine residue (G305) in GLYT1b and a serine residue (S481) in GlyT2a. The G305 in GLYT1b was mutated to a serine (GLYT1-G305S) and S481 in GLYT2a was mutated to a glycine (GLYT2-S481G) using site directed mutagenesis (Vandenberg et al., 2007). In contrast to wild type GLYT2a, sarcosine is a substrate of the GLYT2a-S481G mutant. The maximal current (97 ± 2%) and EC_{50} (26.2 ± 1.3 µM) of sarcosine at the mutant receptor are similar to that for wild-type GLYT1b (87 ± 1%, 22 ± 1 µM). The introduction of the corresponding mutation, G305S, into GLYT1b reduced levels of surface expression to approximately 10% of wild type. To overcome this limitation, two additional mutants were generated, GLYT1b-G305A and GLYT2a-S481A. Glycine transport of both mutant receptors is similar to that of wild type. However, incorporation of the S481A substitution into GLYT2a produced a transporter that could transport sarcosine with an efficacy similar to that of GLYT1b (70 ± 3%), but with a reduced affinity (590 ± 50 µM) relative to wild type. Combined, these findings demonstrate that residues at this position are important for sarcosine transport. For GLYT2a, sarcosine can be transported if an alanine or a glycine residue is present at this site, but not if a serine residue is present.

3.4 Mutagenesis to identify transport and channel domains of glutamate transporters

Glutamate transporters have two distinct functions: ion coupled glutamate transport and glutamate-activated chloride channel activity. In 1995, two studies demonstrated that the two functions co-exist in the same protein (Fairman et al., 1995; Wadiche et al., 1995). Glutamate binding is required for activation of the channel, but the direction of Cl⁻ ion flow through the channel domain is uncoupled from the direction of glutamate transport. This raised the question as to how the protein could support the dual functions. In the following section, we will describe how mutagenesis has been used to understand the structural basis for the dual functions of glutamate transporters. This is an interesting example of the complementary nature of mutagenesis and crystallography approaches to understand the functional properties of this class of transporters.

Prior to the determination of the crystal structure of Glt$_{Ph}$, mutagenesis was used to identify residues that may play a role in transporter function. Valine 452 (V452) of EAAT1 is located in the HP2 domain, and the V452C mutation does not alter the functional properties of the transporter. After modification of the V452C mutant with the methanethiosulfonate (MTS) reagent, [2-(trimethylammonium)ethyl] methanethiosulfonate (MTSET), the protein is no longer capable of transporting glutamate; but it still retains the glutamate-activated chloride channel (Ryan & Vandenberg, 2002). This suggests that the two functions are mediated by distinct conformational states of the transporter. In a separate study, our group attempted to identify regions of the transporter that form the chloride channel. We focussed our mutagenesis studies on TM2. TM2 contains a number of positively charged residues at the extracellular edge of the helix and a number of uncharged serine and threonine residues in the middle of the helix. We postulated that positive charges at the extracellular edge would attract anions into the channel and the hydrophilic residues within the channel would facilitate anion movement through the channel. To address this hypothesis, we mutated the positively charged residues at the extracellular edge to cysteine residues and probed the reactivity of the cysteine residues to both positively and negatively charged MTS reagents. The negatively charged MTS reagents had faster rates of reactivity than the positively charged MTS reagents, which suggested that the positively charged residues do attract negative charges to the extracellular edge of TM2. Substitutions of the hydrophilic serine and threonine residues in the middle of TM2 to small aliphatic residues significantly altered the anion permeability of the channel without affecting the transport function. This confirmed that the two functional properties of the transporter are mediated by separate domains and also that the serine and threonine residues are likely to line the pore of the channel.

The studies described above were carried out prior to any knowledge of the three dimensional structure of the transporter, and so we attempted to identify how close the channel domain was from various other sites on the transporter by cross-linking experiments. Cysteine residues were introduced within TM2 and then at various other sites of the transporter, including V452 (see above). A disulfide bond formed spontaneously between V452C (in HP2) and Q93C (in TM2), indicating that these two residues must come into close proximity. From this study a crude structural model for these parts of the transporter was developed (Fig. 7A).

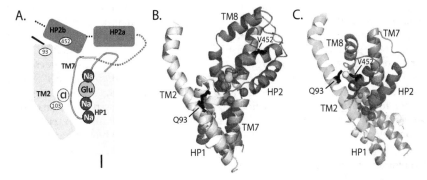

Fig. 7. Structural predictions of the relationship between the glutamate binding and translocation domain and the chloride conducting domain. A. A structural model for glutamate transport and Cl⁻ ion permeation of EAAT1. We have omitted the K⁺ ion and H⁺ for simplicity. The *thick line* is in the plane of the paper and the *dashed line* is behind the plane of the paper. V452C can form a disulfide bond with Q93C. We propose that Cl⁻ ions interact with residues along TM2. In this model, we suggest that glutamate and Na⁺ ions permeate the same pore as Cl⁻ ions, but that there are separate molecular determinants for the two functions. B, C. The structure of the transport domain composed of HP1 (yellow), TM7 (orange), HP2 (red) and TM8 (purple) relative to TM2, which contains molecular determinants for Cl⁻ permeation. Bound aspartate is shown in space-filling representation, and two Na⁺ ions are represented as blue spheres. The residues equivalent to Q93 and V452 are in stick representation and coloured black. B. Shows the distance between Q93 and V452 in the occluded state (PDB 2NWX), while C is the Hg²⁺ cross-linked structure showing the conformational changes required to bring Q93 and V452 into close proximity (PDB 3KBC, Hg²⁺ shown as a yellow sphere).

The crystal structure of Glt$_{Ph}$ was published in late 2004 (see above for a description of the structure); and, whilst the structure revealed many important details about substrate binding, the nature of the mechanism of activation of the chloride channel was not clear. The equivalent residues to V452 and Q93 in Glt$_{Ph}$ are approximately 20 Å apart (Fig. 7B, equivalent residues in a Glt$_{Ph}$ protomer are represented in black), which suggested that these residues were unlikely to come sufficiently close to form a disulfide bond. The first step in resolving this apparent contradiction came from confirming that hydrophilic residues in TM2 of Glt$_{Ph}$ also form part of the lining of the chloride channel lumen, as observed for EAAT1 (Ryan & Mindell, 2007). The laboratory of Olga Boudker then repeated the crosslinking experiments in Glt$_{Ph}$, using cysteine mutants and adopting a similar approach to the one our group had done for EAAT1. Whilst spontaneous disulfide bonds did not form between the two residues in Glt$_{Ph}$, it was possible to catalyse the formation of a bond between the two cysteine residues using Hg^{2+}. It was concluded that the two domains can indeed move sufficiently to allow the two residues to come into close proximity. The cross-linked Glt$_{Ph}$ was also crystallized, and its structure was compared with the original structure (Fig. 7C). This study identified the conformational changes required to bring about the formation of the crosslinks and also suggested a mechanism for the transport process and how this process can lead to channel activation (Reyes et al., 2009). Briefly, the three transport domains (consisting of TM3, TM6, TM7 and TM8 and HP1 and HP2 from each protomer) move as three separate units through a rigid trimerization scaffold. TM2 is part of the scaffold, whilst HP2 is part of the transport domain. It would appear that the sliding movement of the transport domain relative to the rigid trimerization scaffold allows chloride ions to pass through the gap between the two functional domains (Vandenberg et al., 2008). Further mutagenesis will be required to verify this proposal. This series of experiments starting with mutagenesis, followed by crystallography, further mutagenesis and then further crystallography, which will also be followed up by further mutagenesis, highlights how the two approaches to understanding structure and function relationships can complement one another and provoke new ideas and concepts in protein function.

3.5 Substrate affinity and K$^+$ ion coupling in EAAT1

For the last section of this chapter, we will focus on an example of how mutagenesis approaches have been used to understand substrate and ion binding properties of the glutamate transporter family. Many of the residues that have been implicated in substrate and ion binding/translocation (Bendahan et al., 2000; Kavanaugh et al., 1997; Vandenberg et al., 1995) and chloride permeation (Ryan et al., 2004) are conserved throughout the glutamate transporter family. In particular, the carboxy-termini of both the EAATs and Glt$_{Ph}$ are highly conserved and contain the substrate and Na$^+$ binding sites. Despite their significant amino acid identity, the EAATs and Glt$_{Ph}$ display several functional differences. The EAATs transport aspartate and glutamate with similar affinity, while Glt$_{Ph}$ is selective for aspartate over glutamate. In addition, Glt$_{Ph}$ transport is not coupled to the co-transport of H$^+$ or the counter-transport of K$^+$ (Boudker et al., 2007; Ryan et al., 2009). Examination of the amino acid sequences of the substrate binding site of the EAATs and Glt$_{Ph}$ does not reveal any residues that can clearly account for the differences observed in substrate selectivity or affinity. However, an arginine residue is in close proximity to the substrate binding site of both the EAATs and Glt$_{Ph}$, but it is located in TM8 in the EAATs and in HP1 of Glt$_{Ph}$ (Fig 1B). The aim of our study was to investigate the functional effect of the location of a positively charged arginine residue in two members of the glutamate transporter family, EAAT1 and Glt$_{Ph}$.

In order to examine the role of this arginine residue, two double mutant transporters were produced. In EAAT1 the arginine residue was moved from TM8 to HP1 (EAAT1S363R/R477M), and the reverse double mutation was introduced into the gene for Glt$_{Ph}$ (Glt$_{Ph}$R276S/M395R). Switching the arginine residue from TM8 to HP1 in EAAT1 had no effect on substrate selectivity, but it did increase affinity for both glutamate and aspartate and abolished K^+ coupling. The apparent affinity for both L-glutamate and L-aspartate was increased ~130 fold, and it was similar to the affinity of L-aspartate in Glt_{Ph} (Ryan et al., 2009). The counter-transport of one K^+ ion per transport cycle is thought to be important for the relocation of the EAATs to the outward-facing state. The movement of an arginine residue from TM8 to HP1 has potentially slowed the return of the empty transporter to the extracellular facing side, thus contributing to the decrease in observed affinity values. In contrast, the inverse changes in Glt_{Ph} (Glt_{Ph}R276S/M395R) resulted in a functional transporter that has a ~4-fold reduction in the affinity for aspartate compared to wild type. The substitutions did not affect substrate selectivity or introduce K^+ dependence.

The crystal structure of Glt_{Ph} reveals that the backbone carbonyl group of the arginine residue in HP1 forms a direct contact with the substrate. However, our mutagenesis studies suggest that it is the side chain that is influencing transport properties. A possible explanation is that the conformation of the HP1 loop region and also the proximal TM8 is influenced by the arginine side chain and neighbouring residues. Thus, mutating this arginine may influence the conformation of the backbone carbonyl group, which in turn may influence substrate affinity.

The movement of K^+ ions through the transporter is likely to rely upon multiple conformational changes and interactions. Disruption of any of these interactions via a mutation is liable to result in loss of K^+ coupling. However, to introduce K^+ coupling will require multiple mutations. This may explain why the double mutation is sufficient to abolish K^+ coupling in EAAT1, but not introduce it into Glt_{Ph}.

4. Conclusion

Directed mutagenesis has been particularly useful in understanding the structure and function of mammalian membrane proteins. In this chapter we have outlined our approach to structure-activity studies of neurotransmitter transporters in the solute carrier (SLC) families, SLC1 and SLC6. We have used examples from our work on the excitatory amino acid transporters (EAATs), the archaeal aspartate transporter (Glt_{Ph}), glycine transporters (GLYTs) and the prokaryotic leucine transporter ($LeuT_{Aa}$).

5. Acknowledgments

We are grateful for the technical assistance of Audra McKinzie, Cheryl Handford and Marietta Salim. Our research group is funded by the Australian National Health and Medical Research Council and the Australian Research Council.

6. References

Aragon C. & Lopez-Corcuera B. (2003). Structure, function and regulation of glycine neurotransporters. *European Journal of Pharmacology* 479(1-3): 249-262.

Arriza J., Kavanaugh M., Fairman W., Wu Y., Murdoch G., North R. & Amara S. (1993). Cloning and expression of a human neutral amino acid transporter with structural similarity to the glutamate transporter gene family. *Journal of Biological Chemistry* 268(21): 15329.

Aubrey K.R. & Vandenberg R.J. (2001). N[3-(4'-fluorophenyl)-3-(4'-phenylphenoxy)propyl]sarcosine (NFPS) is a selective persistent inhibitor of glycine transport. *British Journal of Pharmacology* 134(7): 1429-1436.

Bartholomaus I., Milan-Lobo L., Nicke A., Dutertre S., Hastrup H., Jha A., Gether U., Sitte H.H., Betz H. & Eulenburg V. (2008). Glycine transporter dimers: evidence for occurrence in the plasma membrane. *Journal of Biological Chemistry* 283(16): 10978-10991.

Bendahan A., Armon A., Madani N., Kavanaugh M. & Kanner B. (2000). Arginine 447 plays a pivotal role in substrate interactions in a neuronal glutamate transporter. *Journal of Biological Chemistry* 275(48): 37436.

Boudker O., Ryan R., Yernool D., Shimamoto K. & Gouaux E. (2007). Coupling substrate and ion binding to extracellular gate of a sodium-dependent aspartate transporter. *Nature* 445(7126): 387-393.

Dohi T., Morita K., Kitayama T., Motoyama N. & Morioka N. (2009). Glycine transporter inhibitors as a novel drug discovery strategy for neuropathic pain. *Pharmacology and Therapeutics* 123(1): 54-79.

Edington A.R., McKinzie A.A., Reynolds A.J., Kassiou M., Ryan R.M. & Vandenberg R.J. (2009). Extracellular loops 2 and 4 of GLYT2 are required for N-arachidonylglycine inhibition of glycine transport. *Journal of Biological Chemistry* 284(52): 36424-36430.

Fairman W.A., Vandenberg R.J., Arriza J.L., Kavanaugh M.P. & Amara S.G. (1995). An excitatory amino-acid transporter with properties of a ligand-gated chloride channel. *Nature* 375(6532): 599-603.

Guastella J., Nelson N., Nelson H., Czyzyk L., Keynan S., Miedel M.C., Davidson N., Lester H.A. & Kanner B.I. (1990). Cloning and expression of a rat brain GABA transporter. *Science* 249(4974): 1303-1306.

Horiuchi M., Nicke A., Gomeza J., Aschrafi A., Schmalzing G. & Betz H. (2001). Surface-localized glycine transporters 1 and 2 function as monomeric proteins in Xenopus oocytes. *Proc Natl Acad Sci U S A* 98(4): 1448-1453.

Ju P., Aubrey K.R. & Vandenberg R.J. (2004). Zn2+ inhibits glycine transport by glycine transporter subtype 1b. *Journal of Biological Chemistry* 279(22): 22983-22991.

Kanai Y. & Hediger M. (2004). The glutamate/neutral amino acid transporter family SLC1: molecular, physiological and pharmacological aspects. *Pflügers Archiv European Journal of Physiology* 447(5): 469-479.

Kavanaugh M., Bendahan A., Zerangue N., Zhang Y. & Kanner B. (1997). Mutation of an amino acid residue influencing potassium coupling in the glutamate transporter GLT-1 induces obligate exchange. *Journal of Biological Chemistry* 272(3): 1703.

Lopez-Corcuera B., Alcantara R., Vazquez J. & Aragon C. (1993). Hydrodynamic properties and immunological identification of the sodium- and chloride-coupled glycine transporter. *J Biol Chem* 268(3): 2239-2243.

Reyes N., Ginter C. & Boudker O. (2009). Transport mechanism of a bacterial homologue of glutamate transporters. *Nature* 462(7275): 880-885.

Ryan R.M., Compton E.L. & Mindell J.A. (2009). Functional characterization of a Na+-dependent aspartate transporter from Pyrococcus horikoshii. *Journal of Biological Chemistry* 284(26): 17540-17548.

Ryan R.M. & Mindell J.A. (2007). The uncoupled chloride conductance of a bacterial glutamate transporter homolog. *Nature Structural and Molecular Biology* 14(5): 365-371.

Ryan R.M., Mitrovic A.D. & Vandenberg R.J. (2004). The chloride permeation pathway of a glutamate transporter and its proximity to the glutamate translocation pathway. *Journal of Biological Chemistry* 279(20): 20742-20751.

Ryan R.M. & Vandenberg R.J. (2002). Distinct conformational states mediate the transport and anion channel properties of the glutamate transporter EAAT-1. *Journal of Biological Chemistry* 277(16): 13494-13500.

Schrodinger L. (2010). The PyMOL Molecular Graphics System, Version 1.3r1.

Shevchuk N.A., Bryksin A.V., Nusinovich Y.A., Cabello F.C., Sutherland M. & Ladisch S. (2004). Construction of long DNA molecules using long PCR-based fusion of several fragments simultaneously. *Nucleic Acids Research* 32(2): e19.

Singh S.K., Piscitelli C.L., Yamashita A. & Gouaux E. (2008). A competitive inhibitor traps LeuT in an open-to-out conformation. *Science* 322(5908): 1655-1661.

Singh S.K., Yamashita A. & Gouaux E. (2007). Antidepressant binding site in a bacterial homologue of neurotransmitter transporters. *Nature* 448(7156): 952-956.

Slotboom D.J., Konings W.N. & Lolkema J.S. (1999). Structural features of the glutamate transporter family. *Microbiology and Molecular Biology Reviews* 63(2): 293-307.

Succar R., Mitchell V.A. & Vaughan C.W. (2007). Actions of N-arachidonyl-glycine in a rat inflammatory pain model. *Molecular Pain* 3: 24.

Supplisson S. & Bergman C. (1997). Control of NMDA receptor activation by a glycine transporter co-expressed in Xenopus oocytes. *Journal of Neuroscience* 17(12): 4580-4590.

Sur C. & Kinney G.G. (2004). The therapeutic potential of glycine transporter-1 inhibitors. *Expert Opinion on Investigational Drugs* 13(5): 515-521.

Vandenberg R.J., Arriza J.L., Amara S.G. & Kavanaugh M.P. (1995). Constitutive ion fluxes and substrate binding domains of human glutamate transporters. *Journal of Biological Chemistry* 270(30): 17668-17671.

Vandenberg R.J., Huang S. & Ryan R.M. (2008). Slips, leaks and channels in glutamate transporters. *Channels (Austin)* 2(1): 51-58.

Vandenberg R.J., Shaddick K. & Ju P. (2007). Molecular basis for substrate discrimination by glycine transporters. *Journal of Biological Chemistry* 282(19): 14447-14453.

Vuong L.A., Mitchell V.A. & Vaughan C.W. (2008). Actions of N-arachidonyl-glycine in a rat neuropathic pain model. *Neuropharmacology* 54(1): 189-193.

Wadiche J.I., Amara S.G. & Kavanaugh M.P. (1995). Ion fluxes associated with excitatory amino acid transport. *Neuron* 15(3): 721-728.

Wiles A.L., Pearlman R.J., Rosvall M., Aubrey K.R. & Vandenberg R.J. (2006). N-Arachidonyl-glycine inhibits the glycine transporter, GLYT2a. *Journal of Neurochemistry* 99(3): 781-786.

Yamashita A., Singh S.K., Kawate T., Jin Y. & Gouaux E. (2005). Crystal structure of a bacterial homologue of Na+/Cl--dependent neurotransmitter transporters. *Nature* 437(7056): 215-223.

Yernool D., Boudker O., Jin Y. & Gouaux E. (2004). Structure of a glutamate transporter homologue from Pyrococcus horikoshii. *Nature* 431(7010): 811-818.

Zerangue N. & Kavanaugh M. (1996). Flux coupling in a neuronal glutamate transporter. *Nature* 383(6601): 634-637.

9

Site-Directed Mutagenesis as a Tool for Unveiling Mechanisms of Bacterial Tellurite Resistance

José Manuel Pérez-Donoso[1] and Claudio C. Vásquez[2]

[1]Universidad de Chile
[2]Universidad de Santiago de Chile
Chile

1. Introduction

Tellurium (Te) is a scarce element in the earth's crust and is not essential for living organisms. It is rarely found in the non-toxic, elemental state (Te°); and the soluble oxyanions, tellurite (TeO_3^{2-}) and tellurate (TeO_4^{2-}), are toxic for most forms of life. Tellurite toxicity has been extensively exploited as a selective agent in diverse microbiological culture media.

A few bacterial tellurite resistance mechanisms have been proposed; but the genetic, biochemical and/or physiological bases underlying TeO_3^{2-} resistance are still poorly understood.

One of our strategies to study bacterial resistance to TeO_3^{2-} has been the cloning and characterization of genes from tellurite-resistant bacteria using *Escherichia coli* as a sensitive host. Using this experimental approach, we have previously shown that the genes *cysK*, *iscS* and *cobA* [encoding cysteine synthase, cysteine desulfurase and S-Adenosyl-L-methionine:uroporphirin-III C-methyltransferase (SUMT), respectively] from the thermophilic rod *Geobacillus stearothermophilus* V mediate tellurite resistance when expressed in *E. coli*. All of these genes were subjected to site-directed mutagenesis to demonstrate their participation in tellurite resistance in this mesophilic host (Vásquez et al., 2001; Tantaleán et al., 2003; Araya et al., 2009). More recently we conducted similar mutagenesis experiments with the *Aeromonas caviae* ST *lpdA* gene, encoding dihydrolipoil dehydrogenase, and found that two amino acid residues are involved in the tellurite reductase branch-activity of this enzyme (unpublished data).

Site-directed mutagenesis, also referred to as site-specific or oligonucleotide-directed mutagenesis, is a technique in molecular biology that allows the creation of mutations at a defined DNA sequence. In general, a synthetic primer containing the desired base change is hybridized to a single-stranded DNA containing the gene of interest; the rest of the gene is then copied using a DNA polymerase. The double-stranded DNA molecule thus obtained is ligated to an appropriate vector and introduced into a host cell for mutant selection.

This chapter does not intend to be an extensive review of tellurite resistance. Instead it was written as an example to make young scientists see how simple observations can help to state the basis of much more complex networks underlying a particular, defined phenomenon.

2. The enigma of tellurite toxicity

The ability of bacteria to counteract the effect of heavy metals has interested microbiologists for many years. Toxic heavy metals are often encountered in nature in many different forms. In air, they exist as metal or oxide dust. In surface and ground water, they are found attached to humic substances; and they also bind to soil and sediments.

Tellurium has applications in the semiconductor industry and electronics (in the production of thermoelements, photoelements and other devices in automation equipment). The increasing demand for new and different semiconductors necessitates research work on the application of various tellurium compounds as semiconductor components.

As a group, microorganisms display resistance to nearly all metal and non-metal ions that are considered toxic to the environment, including Ag^+, As^{3+}, Cd^{2+}, Cr^{3+}, Hg^{2+}, Sb^{3+}, Te^{4+}, Te^{6+} and Zn^{2+}, among others (Silver, 2006). Although the literature on the subject is vast and continuously updated (Silver, 2011), in most cases, however, the knowledge of the biochemical and/or genetic mechanisms underlying the metal resistance phenomena is still very limited. This is particularly true for bacterial tellurite resistance, in which much effort has been expended to understand how bacteria counteract the toxic effects of the tellurium salt.

Tellurium (Te) was considered almost an exotic element and was treated with certain indifference by most serious chemists. However, the impressive number of publications on Te compounds during the last few years shows that Te is now widely used in applied chemical reactions.

The natural Te cycle has not been investigated in depth, and the role of microbes – if any - in this process has not yet been elucidated. Nevertheless, tellurite-resistant bacteria do exist in nature; and they often reduce tellurite to its elemental, less toxic, form (Te^o), which accumulates as black deposits inside the cell (Taylor, 1999; Chasteen et al., 2009).

As a result of the accumulated knowledge, several tellurite-resistance determinants (Te^R) have been localized on plasmids and on the chromosome. Structure and organization vary greatly among bacterial species (Taylor, 1999). It has been argued that tellurite toxicity results from the ability of tellurite to act as a strong oxidizing agent that damages a number of cell components (Taylor, 1999; Pérez et al., 2007). In the last years, however, available evidence shows that tellurite toxicity results from the generation of reactive oxygen species (ROS) (Borsetti et al., 2005; Calderón et al., 2006; Tremaroli et al., 2007; Pérez et al., 2007). ROS, such as hydrogen peroxide (H_2O_2), superoxide anion (O_2^-) and hydroxyl radical ($OH^·$), are typical byproducts of aerobic metabolism. However, they can also be produced upon exposure of the cell to free radical-generating compounds, like metals and metalloids.

Our group has been interested in studying tellurite resistance (Te^R)/toxicity for many years. First we focused on thermophilic, Gram-negative, rods of the genus *Thermus* and later on *G. stearothermophilus* V, a thermotolerant, spore-forming, Gram-positive bacterium that was isolated in our laboratory from a soil sample. In both cases we demonstrated the existence of cellular reductases that convert tellurite into elemental tellurium at the expense of NAD(P)H oxidation *in vitro* (Chiong et al., 1988; Moscoso et al., 1998).

In an attempt to identify genetic determinants for Te[R] in these bacteria, we constructed gene libraries that were used to transform sensitive *E. coli* hosts to tellurite resistance. While the cloning of resistance determinants from *Thermus* has been unsuccesful so far, we did clone tellurite resistance determinants from *G. stearothermophilus* V into *E. coli*. These genes were subjected to site-directed mutagenesis in order to unveil their participation in the resistance phenomenon (Vásquez et al., 2001; Tantaleán et al., 2003; Araya et al., 2009). More recently, we have shown that overproduction of the *Aeromonas caviae* ST dihydrolipoil dehydrogenase results in enhanced tellurite resistance in *E. coli*. This enzyme exhibits NADH-dependent tellurite reductase (TR) activity (Castro et al., 2008, 2009). The change of two defined amino acid residues at the enzyme active site decreased TR activity (unpublished data). What follows is a chronological description of the above-mentioned results.

2.1 CysK

Cysteine synthases (CysKs) are enzymes that catalyze the last step in cysteine biosynthesis. They have been related to tellurite resistance in different microorganisms (Moore & Kaplan, 1992; O'Gara et al., 1997; Alonso et al., 2000; Vásquez et al., 2001; Lithgow et al., 2004). All cysteine synthases described to date require the cofactor pyridoxal 5'-phosphate (PLP) for activity. PLP-dependent enzymes catalyze a broad spectrum of aminoacid transformations involved in the development of an organism, such as transaminations, β-eliminations, β-γ replacements and racemizations.

Searching for tellurite-resistance determinants, we identified and characterized a new thermophilic CysK in *G. stearothermophilus* V (Saavedra et al., 2004). *E. coli* ovexpressing this *cysK* gene shows a tellurite resistance that is over 10-fold more than that observed for the wild-type controls. Despite the fact that it is known that the ping-pong catalytic mechanism of this enzyme is similar to that of other CysKs, the *G. stearothermophilus* V enzyme has not been fully characterized. The catalytic amino acid residues are also not known. In addition, the importance of cysteine synthase in bacterial tellurite resistance has not been totally documented; and our group has proposed that the resistance increase is paralleled by increased levels of reduced thiols, such as glutathione.

G. stearothermophilus V CysK is a homodimer (32 kDa/monomer) that requires one PLP molecule per subunit (Saavedra et al., 2004). It belongs to the β family of PLP-dependent enzymes and shares some similarities with other enzymes involved in deamination reactions, as do tryptophan synthase, threonine deaminase and O-acetyl-serine sulphydrilase (Alexander et al., 1994).

A general mechanism for the CysK-catalyzed reaction has been proposed (Cook and Wedding, 1977; Tai et al., 1998). The enzyme binds PLP by a lysine group forming a Schiff base, known as internal aldimine. This intermediate absorbs in the 400-430 nm region and exhibits two resonant forms. Addition of the O-acetyl-L-serine (OAS) substrate allows the formation of the geminaldiamine intermediate, which produces the external aldimine. Then the quinonoid intermediate is formed; and when the substituent in the β position is released (acetate), the α-aminoacrylate is finally formed.

To study the residues and motifs that define the catalytic properties of this enzyme, we used site-directed mutagenesis to assess the importance of the C-terminus and of some putative catalytic residues for CysK activity and CysK-mediated bacterial tellurite resistance (unpublished data).

As a first approach, a set of CysK C-terminal deletions of 10 (CysK ΔTyr298), 20 (CysK ΔLeu288), 30 (CysK ΔAla278), 40 (CysK ΔGly268) and 60 (CysK ΔAla248) amino acids were constructed, overexpressed, purified and characterized. Binding of the PLP cofactor was evaluated through the absorption spectrum of the purified proteins. An absorbance peak at 412 nm is characteristic of the α-aminoacrylate intermediate.

All the CysK deletion mutants larger than ΔTyr298 were inactive, unable to bind PLP and did not confer tellurite resistance. This result indicated a direct relationship between enzymatic activity and tellurite tolerance. It is also in agreement with a role for thiols, such as cysteine, in tellurite tolerance. In this context, increased levels of intracellular reduced thiols, particularly glutathione, were observed in cells overproducing CysK. This observation suggested that increased concentrations of cell antioxidants could be responsible for protecting *cysK*-overexpressing cells from tellurite-mediated oxidation. Interestingly, the mutant having Tyr298 (Tyrosine as residue 298, Y298) (see Table 1 in the chapter by Figurski et al. for the amino acid codes) as the C-terminal residue [CysK Δ(Tyr298)] was inactive, despite retaining the ability to bind PLP and to form the α-aminoacrylate intermediate. As expected, overproduction of this mutant enzyme did not enhance tellurite resistance in *E. coli*.

As shown in Fig. 1, CysK displays the conserved amino acid sequence motif SVKDRIA near the amino terminus, which is required for PLP binding. Most of the C-terminal truncated mutants were unable to bind PLP, despite the presence of this motif and the finding that the proteins were correctly translated and folded in the cytoplasm. This observation suggests that other residues are involved in stabilizing PLP binding. The residues are probably located near the CysK C-terminus.

As can be deduced from these results, protein deletions allow a global view about the importance of some protein motifs and not a detailed interpretation on the participation of defined amino acids in the enzyme's functioning. In this context, site-directed mutagenesis offers a more versatile alternative to study the role of a defined motif or amino acid residue. These two experimental approaches can complement each other in order to obtain a detailed analysis for understanding the enzymatic mechanism.

To choose the appropriate residues to be subjected to site-directed mutagenesis, the first approach involved sequence conservation studies using BLASTP and ClustalW software and other programs. However, a better idea about which domains and/or residues could be interacting with defined molecules or atoms in a reaction can be obtained by constructing 3D models of the protein. In those cases where there is no crystallographic information regarding the protein, bioinformatic tools can offer a useful alternative in order to predict a model based on sequence homology.

Sequence analysis allowed identification of the conserved [43]SVKDRIA[49] domain (NCBI accession number AAG28533.1, see Fig. 1) in the *G. stearothermophilus* V CysK. In this conserved sequence, K45 is required to form the protonated Schiff base with PLP. K45 also

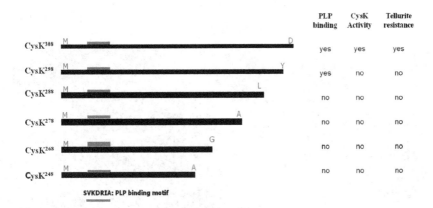

Fig. 1. Schematic model representing the different truncated CysKs and their properties (see the text for details).

participates directly in binding the cofactor and NH_3 transamination. A replacement of K45 by alanine (K45A) was made in CysK to evaluate the importance of the positive charge required for PLP interaction. The mutated gene was overexpressed in *E. coli*, and the enzyme was purified to homogeneity. No enzymatic activity or cofactor binding was observed by the purified enzyme, confirming that this residue is required for PLP attachment, as predicted.

No structural data are available for *G. stearothermophilus* V CysK, so a 3D model based on the available crystallographic structure (2 A° resolution) for the *Salmonella enterica* serovar Typhimurium CysK was constructed. With this model, we expected to make predictions about residues located at the enzyme's C-terminus that could be interacting with PLP, stabilizing it and allowing the transamination reaction.

As mentioned before, the deletion mutant CysK Δ(Tyr298) was inactive, but still able to bind PLP. Based on this observation, a homology model of this deletion mutant was constructed to define the position of specific residues that can participate in PLP stabilization both in wild-type CysK and in the CysK Δ(Tyr298) mutant.

The model suggested that the CysK Δ(Tyr298) deletion mutant exhibits amino acids in positions that can interact with PLP so as to orientate the cofactor for the reaction to proceed (Fig. 2). Tyrosine 298 is most probably interacting directly with PLP, given that it is located at a distance (less than 4 A°) that allows the formation of a hydrogen bond with the cofactor. Experiments with this deletion mutant confirmed that the enzyme was able to bind PLP and was folded correctly. However, studies performed to assess PLP- mediated fluorescence indicated that the cofactor displayed a different orientation than that in wild-type enzyme. Altogether, these results suggested that the Tyr298 residue could be important for CysK activity. Most probably Tyr298 forms a hydrogen bond through the hydroxyl group of the tyrosine, thus stabilizing and favoring a correct orientation of PLP. To test this possibility, homology analyses and new models of Y298A and Y298P CysK mutants were constructed to predict the role of the Tyr298 residue. The results suggested that the -OH group of Tyr298 is required for PLP binding. It is thought to affect catalysis because of an interaction with the α–aminoacrylate intermediate through a water molecule.

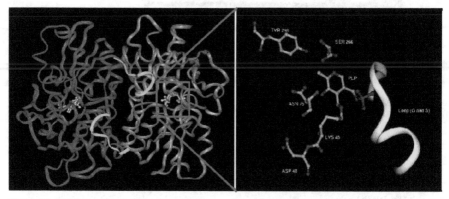

Fig. 2. Model (homology) of the *G. stearothermophilus* V CysK. Left, CysK homodimer showing one PLP molecule (yellow) bound per subunit and the C-terminal 10 amino acids (green). Right, schematic view of those amino acid residues involved in PLP binding (LYS45) and in orienting and/or stabilizing the cofactor (SER266, TYR298, and ASN75).

Two additional Tyr298 site-directed variants were constructed and characterized (Y298A and Y298F). Mutant proteins were purified and assayed for PLP binding, α-aminoacrylate intermediate formation, and enzymatic activity (lyase and cysteine synthase). Both mutant enzymes were inactive and did not bind PLP nor formed the α-aminoacrylate intermediate, suggesting that the -OH group of Tyr298 is required for CysK activity.

Based on these observations, we propose that the Tyr298 residue of *G. stearothermophilus* V CysK is involved in stabilization and proper orientation of PLP by interacting with the phosphate group of the cofactor through hydrogen bonding. The expression of the *G. stearothermophilus* V *cysK* gene was later shown to mediate low-level tellurite resistance in *E. coli* (Vásquez et al., 2001).

2.2 IscS

One of the superoxide anion targets is a family of dehydratases that use exposed [4Fe-4S] clusters to bind and dehydrate their substrates during the biosynthesis of branched-chain amino acids. Oxidation of these proteins results in the dismantling of these [Fe-S] centers with the concomitant loss of enzyme activity. Crude *E. coli* extracts catalyzing the formation of these centers *in vitro* contain at least four enzymatic activities that provide the required sulfur atoms: O-acetyl-serine sulfhydrilase A (CysK), O-acetyl-serine sulfhydrilase B (CysM) and β-cistationase; the fourth protein displayed cysteine desulfurase activity. Cysteine desulfurases remove the sulfur atom from cysteine to construct and repair [Fe-S] clusters in protein substrates that, in turn, catalyze essential redox reactions in critical metabolic pathways. Very soon it became clear that the IscS cysteine desulfurase played an important role in transferring the sulfur from cysteine for [Fe-S] center synthesis *in vivo*.

Various tellurite-resistant (Te^R) *E. coli* clones were isolated upon transformation with a *G. stearothermophilus* V *Hind*III library. In particular, one contained a 3.5 kilobase (kb) DNA insert that specified three open reading frames (ORFs). By comparison with sequences deposited in protein data banks, it was found that ORF2 [1200 base pairs (bp)] encoded a ~45 kDa cysteine desulfurase (IscS). The expression of the *G. stearothermophilus* V IscS cysteine desulfurase conferred tellurite resistance in *E. coli*. The enzyme was induced and purified to homogeneity; the purified enzyme displayed cysteine desulfurase activity. We showed that tellurite resistance depends in part on the activity of the IscS enzyme, supporting the hypothesis that essential proteins with iron-sulfur [Fe-S] clusters are among the main targets of the oxidative damage caused by tellurite in *E. coli*. Unlike the case for other microbes, the *G. stearothermophilus* V *iscS* gene does not appear to be within an operon containing other genes involved in *de novo* [Fe-S] cluster formation.

Because *G. stearothermophilus* V is a thermophile, IscS purification included an incubation of soluble cell extracts at 70 °C for 20 min. This step eliminated almost 75% of the total starting protein, without an appreciable loss of IscS protein or cysteine desulfurase activity (Fig. 3, lane 1). Subsequent column chromatography steps resulted in enzyme preparations that were >98% pure. The amino-terminus of the purified IscS (MNLEQIRKDTPLHKKYSYIN), determined by Edman degradation, matched precisely the predicted primary sequence of the product of the *iscS* gene. The native form of the enzyme is a homodimer with an apparent molecular mass of 93-97 kDa, as determined by size-exclusion chromatography. This cysteine desulfurase belongs to the α-family of PLP-dependent enzymes and exhibits an absorbance maximum for PLP centered at about 420 nm, the characteristic UV-visible spectrum of other cysteine desulfurases.

To further confirm that IscS activity was responsible for tellurite resistance in *E. coli*, a series of mutant derivatives was constructed. Plasmids containing truncated versions of the *iscS* gene (90, 150 or 210 bp deletions of the 3′ end) did not confer resistance to tellurite in *E. coli*; nor did they exhibit desulfurase activity in crude extracts. The induced mutant proteins formed inclusion bodies, suggesting that the carboxyl terminus of IscS is essential for proper folding, dimerization and/or function.

We decided to make a directed change of Lys213, a residue that likely binds the PLP cofactor. This lysine is conserved in the sequence of cysteine desulfurases from both Gram-positive and Gram-negative bacteria, as well as in the yeast *Saccharomyces cerevisiae*. Lys213 in IscS was replaced by alanine to yield the *iscS_K213A* mutant gene, which was cloned into the expression plasmid pET21b and introduced into *E. coli* JM109(DE3) to induce high-level transcription from the promoter. Cells expressing *iscS K213A* did not exhibit tellurite resistance.

The K213A enzyme was purified to homogeneity, using the same procedure as for the native enzyme, but omitting the initial heat-treatment because the mutant protein did not show the thermostability of the wild-type IscS (Fig. 3). Unlike extracts containing IscS, those of the IscS K213A mutant protein did not exhibit the typical intense yellow colour, consistent with the idea that Lys213 is critical for PLP binding. In fact, the absorbance peak characteristic of PLP-containing enzymes was missing in the UV-visible spectrum of the purified mutant protein. The mutant IscS enzyme showed less than 10% of the specific activity exhibited by the wild-type IscS.

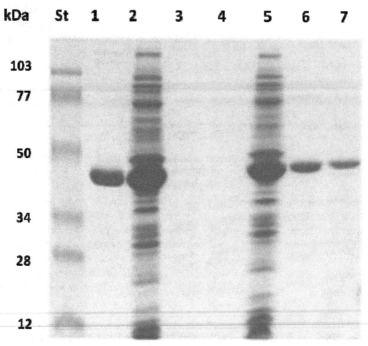

Fig. 3. Effect of temperature on the IscS K213A and K213R mutants. Lane 1, wild-type IscS. Lanes 2, 3 and 4, crude extracts of *E. coli* overproducing the K213A mutant cysteine desulfurase incubated for 10 min at 37 °C, 10 min at 70 °C and 20 min at 70 °C, respectively. Lanes 5, 6, and 7, as in lines 2-4, but using *E. coli* extracts overproducing the K213R mutant cysteine desulfurase.

On the other hand, an IscS K223R mutant was also constructed to confirm the importance of the positive charge in protein stabilization, PLP binding and desulfurase activity. As shown in Figs. 3 and 4, the K213R IscS displayed the same thermostability behavior of the wild-type desulfurase; and the enzyme was also easily purified by heating. In addition, the purified K213R mutant displayed the characteristic intense yellow color of PLP enzymes, an observation that was further confirmed by the presence of the 412 nm absorbance peak in the UV-visible spectra. As expected, this mutant displayed desulfurase activity levels identical to those of the wild-type desulfurase (Fig. 4). These results confirmed that the arginine residue was able to maintain IscS function, indicating that the positive charge at this position is required for proper PLP binding (Fig. 5).

The *G. stearothermophilus* V *iscS* gene not only complemented successfully an *E. coli iscS* mutation, but also conferred tellurite resistance to an *E. coli sodAsodB* double mutant, arguing that superoxide causes specific damage to one or more critical [Fe-S] cluster-containing proteins.

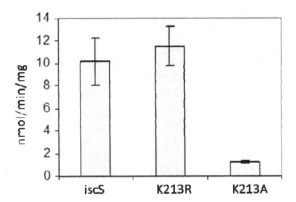

Fig. 4. Desulfurase activity of wild-type IscS and the indicated mutant IscSs.

2.3 CobA

A third gene mediating tellurite resistance in *E. coli* was identified from a *G. stearothermophilus* V HindIII library. The *cobA* gene was initially identified as one of three main ORFs present in a 3.8 kb *G. stearothermophilus* V DNA insert in the recombinant plasmid p1VH. *E. coli* carrying p1VH exhibited over 10 fold the resistance to potassium tellurite observed in the same strain harboring the pSP72 cloning vector alone (Araya et al., 2004).

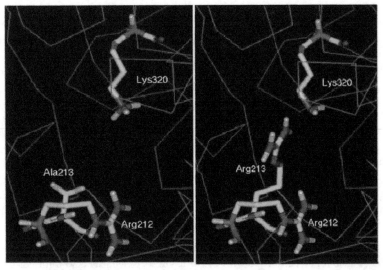

Fig. 5. Molecular models of the active site of the K213A and K213R mutants of the IscS enzyme from *G. stearothermophilus* V.

The *cobA* gene of *G. stearothermophilus* V encodes a 28 kDa protein exhibiting 71% identity with the *Bacillus megaterium cobA* gene, which encodes the enzyme S-adenosyl-L-methionine:uroporphirin-III C-methyltransferase, also referred to as SUMT.

The *cobA* gene was amplified by PCR using appropriate primers, inserted into the pET21b expression vector and introduced into *E. coli* JM109(DE3). Transformants exhibited higher tellurite resistance than that observed for cells carrying the cloning vector alone. Minimal inhibitory concentration (MIC) determinations were carried out in the absence of the inducer (IPTG), since the protein seemed to be toxic to the cell when overexpressed.

G. stearothermophilus V SUMT was induced with IPTG and judged >95% pure after being purified in two chromatographic steps (Cibacron blue and Sephadex column chromatography). After being fractionated by polyacrylamide gel electrophoresis in the presence of sodium dodecyl sulphate (PAGE-SDS) and transferred to a polyvinylidene fluoride (PVDF) membrane, the purified SUMT was sent for EDMAN microsequencing analysis. The first 10 amino acids (MTNGKVYIVG) matched 100% of those predicted by the nucleotide sequence of the *cobA* gene. A Mr of 60 kDa, compatible with a homodimeric quaternary structure, was deduced for the SUMT enzyme by electrophoresis under non-denaturing conditions and by gel chromatography.

The *cobA*-encoded amino acid sequence was compared with that of other methylases involved in corrinoid biosynthesis. Highly conserved regions were present in all analyzed sequences. One of them, whose consensus is GXGXGD, has been described as a SAM-binding motif (Fig. 6), suggesting that the *G. stearothermophilus* V *cobA* gene product is actually an uroporphirinogen III-like C-methyltransferase. Appropriate primers were designed to introduce a change of A to G at position 12 of the SUMT enzyme by recombinant PCR. The product was cloned into the pET21b(+) expression vector and the A12G mutant protein was purified as above. As the wild-type counterpart, the SUMT A12G protein exhibited a homodimeric structure, as determined by non-denaturing polyacrylamide gel electrophoresis and size exclusion chromatography. Cells expressing the mutant SUMT showed low K_2TeO_3 resistance (MIC 2.5 µg/ml) as compared to the wild-type clone (MIC 18 µg/ml).

Various deletion mutants of the *cobA* gene were constructed by PCR that included 60-, 120- and 180-bp deletions from its 3' end. The truncated DNA fragments were amplified and cloned into the pET21b(+) expression vector. Plasmids carrying *cobA*Δ60, *cobA*Δ120 and *cobA*Δ180 were introduced into *E. coli* JM109(DE3) by transformation. Although expressed in high amounts, purification of the truncated proteins failed, as they formed inclusion bodies. Several attempts to solubilize them were carried out also without results. All clones expressing truncated genes were sensitive to tellurite (MIC 1.25 µg/ml).

Addition of methyl-³H SAM to the purified SUMT enzyme followed by size exclusion chromatography revealed two radioactive peaks, corresponding to enzyme-bound and free SAM. When the wild-type enzyme was replaced by the SUMT A12G mutant in the SAM binding assay, only the free SAM peak was observed.

Another strategy of a cell to cope with the toxic effects of tellurite is to form volatile, less toxic, compounds with it. In this context, headspace analysis from cultures of different bacteria by gas chromatography- fluorine induced chemiluminescence detection (GC-

F2ICD) has proven to be useful for detecting the evolution of sulfur compounds, such as methanethiol (MeSH), dimethyl sulfide (DMS), dimethyl disulfide (DMDS), dimethyl trisulfide (DMTS), and organotellurides, like dimethyl telluride (DMTe) (Chasteen and Bentley, 2003).

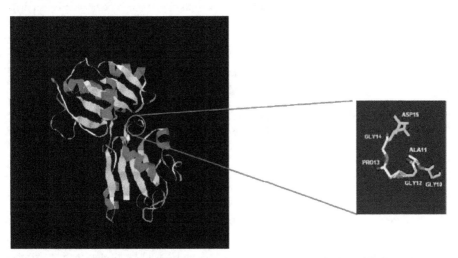

Fig. 6. Model (homology) of the SUMT methyl transferase. The inset shows the SAM-binding motif.

Given that SUMT is a methyltransferase, it was tempting to correlate it *a priori* with the resulting volatile Te derivatives. Since methylcobalamins (methyl-B12) have been involved in biomethylating a number of heavy metals and metalloids, one putative way by which SUMT could participate in tellurite resistance in *E. coli* could be precisely through this kind of methylation. However, two lines of evidence allow this assumption to be discarded. First, *E. coli* does not synthesize methyl-B12 *de novo*; and, second, amending the culture medium with 10 mM CoCl$_2$ did not change the K$_2$TeO$_3$ MIC. It is well known that Co salts inhibit any methyl-B12-mediated methylation. Thus, a putative role of SUMT in K$_2$TeO$_3$ resistance would be the utilization of Te as a substrate and to catalyze the transfer of methyl groups from SAM to the metalloid. In this context, some work from other authors has indicated that the gene products of the *tpm* and *tehB* genes from *Pseudomonas syringae* and *E. coli*, respectively, are able to biomethylate tellurium (Cournoyer et al., 1998; Liu et al., 2000). Unfortunately, we were unable to detect the genesis of methylated tellurium derivatives in the headspace of cells cultured in the presence of tellurite or tellurate. To date there is no clear experimental evidence that sheds light on the enzymatic mechanism underlying tellurium biomethylation.

On the other hand, since sulfite reductase (reduces sulfite to sulfur) utilizes siroheme as a prosthetic group, SUMT could participate in tellurite tolerance by enhancing the biosynthesis of this cofactor and, hence, that of cysteine. In the same context, it was found that enzymes that reduce thiols (glutathione and thioredoxin reductases) and their metabolites (thioredoxins, glutaredoxins and glutathione) would be involved in tellurite resistance. Recent results from our laboratory indicate that when grown in the presence the toxicant, the total thiol content is higher in cells expressing the *cobA* gene than in cells carrying the vector alone. Our interpretation is that SUMT could participate in the generation of reducing power (cysteine, for example) that would be used to compensate (or to recover) GSH or another metabolically important thiol that could have been consumed during tellurite reduction.

2.4 LpdA

As mentioned before, one of the most relevant properties of potassium tellurite is its high toxicity for microorganisms. In this context, our approach to understand the basis of the toxic effects has been the search of resistance determinants in tellurite-resistant strains, such as *G. staerothermpophilus* V. Following the same idea, another highly TeO$_3^{2-}$-resistant bacterial strain was isolated from environmental water. This new strain exhibited a tellurite MIC close to 300 µg/ml. It was identified as the Gram-negative *Aeromonas caviae* ST. In addition to its high resistance to tellurite, this strain exhibited high levels of tellurite reduction, as determined by the darkness of cells exposed to the toxic salt and by tellurite reductase (TR) enzymatic assays performed with cell-free extracts. Interestingly, most of this TR activity was dependent of NADH and tracked to the pyruvate dehydrogenase multienzymatic complex (PDH), specifically to the E3 component encoded by the *lpdA* gene (Castro et al., 2008, 2009).

The *lpdA* gene was cloned; and the recombinant plasmid was used as template to construct three different mutants by site-directed mutagenesis: C45A, H322Y and E354K (Fig. 7). These mutants were chosen based on previous work on the *E. coli* E3 component indicating that C45 is highly conserved and is involved in the formation of a disulfide bond with C50, required for appropriate protein conformation (Kim et al., 2008). H322 and E354 were changed to Y and K, respectively, because it was previously shown that these mutations affect NADH binding, a substrate required for PDH as well as for TR activity (Castro et al., 2008, 2009).

In this case a different and easier approach to construct the mutants was carried out. Using a high-fidelity and highly processive DNA polymerase and two complementary primers, the plasmid was amplified by PCR and then the methylated template was digested with *Dpn*I restriction endonuclease and used to transform *E. coli*.

As expected, changes of these amino acids resulted in negative effects on pyruvate dehydrogenase activity in cells overproducing these proteins as compared to controls. Decreased PDH activity was observed, particularly in the cases of the H322Y and E354K mutants. The effect was not so pronounced in mutants that do not affect NADH binding. Regarding tellurite reductase activity (TR), an important decrease (~70%) was observed in all three mutants, as determined with purified proteins or in crude extracts of cells overproducing the respective mutant (Fig. 8). These results confirm the importance of NADH for PDH and TR activities and also indicated that C45, while relevant for LpdA-mediated tellurite reduction, is not absolutely required for PDH activity. This idea is in agreement with previous observations of our group and others regarding the importance of cysteine in tellurite resistance (Vásquez et al., 2001; Fuentes et al., 2007).

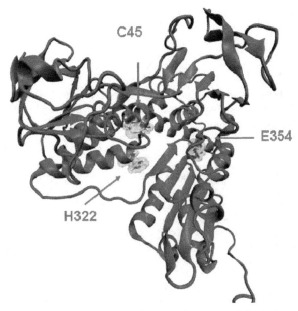

Fig. 7. LpdA model (homology) indicating the spatial position of the amino acids targeted for site-directed mutagenesis.

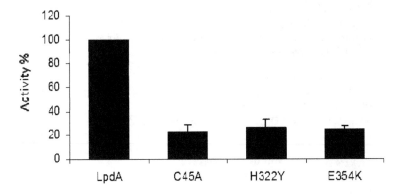

Fig. 8. Tellurite reductase activity of purified LpdA and the indicated LpdA mutants.

3. Conclusion

Several different mechanisms have been proposed to account for the toxicity of tellurite. Tellurium may replace sulfur and/or selenium in critical metabolites or enzymes and abate their essential functions. Alternatively, tellurite is a strong oxidizing agent that may cause general oxidative damage; or it may cause specific damage to critical thiol groups or [Fe-S] clusters present in essential enzymes. The results of this chapter point out different instances in which diverse metabolic pathways, their substrates or products play a still not well-defined role in bacterial tellurite resistance.

4. Acknowledgments

The authors thank Fondecyt grants # 1090097 and 3100049 (Fondo de Desarrollo Científico y Tecnológico, Chile) and Dicyt-USACH (Dirección de Investigación en Ciencia y Tecnología-Universidad de Santiago de Chile).

5. References

Alexander, F.W., Sandmeier, E., Mehta, P.K. & Christen, P. (1994). Evolutionary relationships among pyridoxal-5'-phosphate-dependent enzymes. Regio-specific alpha, beta and gamma families. *Eur. J. Biochem.*, Vol. 219, No. 3, (February 1994), pp. 953-960, ISSN 0014-2956

Alonso, G., Gomes, C., González, C. & Rodríguez-Lemoine, V. (2000). On the mechanism of resistance to channel-forming colicins (PacB) and tellurite, encoded by plasmid Mip233 (IncHI3). *FEMS Microbiol. Lett.*, Vol. 192, No. 2, (November 2000), pp. 257-261, ISSN 0378-1097

Araya, M.A., Tantaleán, J.C., Fuentes, D.E., Pérez, J.M., Calderón, I.L., Saavedra, C.P., Chasteen, T.G. & Vásquez, C.C. (2009). Cloning, purification and characterization of *Geobacillus stearothermophilus* V uroporphirinogen-III C-methyltransferase: evaluation of its role in resistance to potassium tellurite in *Escherichia coli*. *Res. Microbiol.*, Vol. 160, No. 2, (March 2009), pp. 125-133, ISSN 0923-2508

Borsetti, F., Tremaroli,V., Michelacci, F., Borghese, R., Winterstein, C., Daldal, F. &, Zannoni, D. (2005). Tellurite effects on *Rhodobacter capsulatus* cell viability and superoxide dismutase activity under oxidative stress conditions. *Res Microbiol.*, Vol. 156, No. 7, (August 2005), pp. 807-813, ISSN 0923-2508

Calderón, I.L., Arenas, F.A., Pérez, J.M., Fuentes, D.E., Araya, M.A., Saavedra, C.P., Tantaleán, J.C., Pichuantes, S.E., Youderian, P.A. & Vásquez, C.C. (2006). Catalases are NAD(P)H-dependent tellurite reductases. *PLoS ONE*, Vol. 1, No. 1, (December 2006), pp. e70, ISSN 1932-6203

Castro, M.E., Molina, R., Díaz, W., Pichuantes, S.E. & Vásquez, C.C. (2008). The dihydrolipoamide dehydrogenase of *Aeromonas caviae* ST exhibits NADH dependent tellurite reductase activity. *Biochem. Biophys. Res. Commun.*, Vol. 375, No. 1, (October 2008), pp. 91-94, ISSN 0006-291X

Castro, M.E., Molina, R.C., Díaz, W.A., Pradenas, G.A. & Vásquez, C.C. (2009). Expression of *Aeromonas caviae* ST pyruvate dehydrogenase complex components mediate tellurite resistance in *Escherichia coli*. *Biochem. Biophys. Res. Commun.*, Vol. 380, No. 1, (February 2009), pp. 148-152, ISSN 0006-291X

Chasteen, T.G. & Bentley, R. (2003). Biomethylation of selenium and tellurium: microorganisms and plants. *Chem. Rev.*, Vol. 103, No. 1, (January 2003), pp. 1-25, ISSN 0009-2665

Chasteen, T.G., Fuentes, D.E., Tantaleán, J.C. & Vásquez, C.C. (2009). Tellurite: history, oxidative stress and molecular mechanisms of resistance. *FEMS Microbiol. Rev.*, Vol. 33, No. 4, (July 2009), pp. 820-832, ISSN 1574-6976

Chiong, M., Barra, R., González, E. & Vásquez C. (1988). Resistance of *Thermus* spp. to potassium tellurite. *Appl. Environ. Microbiol.*, Vol. 54, No. 2, (February 1988), pp. 610-612, ISSN 0099-2240

Cook, P.F. & Wedding R.T (1977). Overall mechanism and rate equation for O-acetylserine sulfhydrylase. *J. Biol. Chem.*, Vol. 252, No. 10, (May 1977), pp. 3549-3553, ISSN 0021-9258

Cournoyer, B., Watanabe, S. & Vivian, A. (1998). A tellurite-resistance genetic determinant from phytopathogenic pseudomonads encodes a thiopurine methyltransferase: evidence of a widely-conserved family of methyltransferases. *Biochim Biophys Acta.*, Vol. 1397, No. 2, (April 1998), pp. 161-168, ISSN 0304-4165

Fuentes, D.E., Fuentes, E.L., Castro, M.E., Pérez, J.M., Araya, M.A., Chasteen, T.G., Pichuantes, S.E. & Vásquez, C.C. (2007). Cysteine metabolism-related genes and bacterial resistance to potassium tellurite. *J. Bacteriol.*, Vol. 189, No. 24, (October 2007), pp. 8953-8960, ISSN 0021-9193

Kim, Y., Ingram, L. & Shanmugam, T. (2008). Dihydrolipoamide dehydrogenase mutation alters the NADH sensitivity of pyruvate dehydrogenase complex of *Escherichia coli* K-12. *J. Bacteriol.*, Vol. 190, No. 11, (March 2008), pp. 3851-3858, ISSN 0021-9193

Lithgow, J.K., Hayhurst, E.J., Cohen, G., Aharonowitz, Y. & Foster, S.J. (2004). Role of a cysteine synthase in *Staphylococcus aureus*. *J. Bacteriol.* Vol. 186, No. 6, (March 2004), pp. 1579-1590, ISSN 0021-9193

Liu, M., Turner, R.J., Winstone, T.L., Saetre, A., Dyllick-Brenzinger, M., Jickling, G., Tari, L.W., Weiner, J.H. & Taylor, D.E. (2000). *Escherichia coli* TehB requires S-adenosylmethionine as a cofactor to mediate tellurite resistance. *J. Bacteriol.*, Vol. 182, No. 22, (November 2000), pp. 6509-6513, ISSN 0021-9193

Moore, M. & Kaplan, S. (1992). Identification of intrinsic high-level resistance to rare-earth oxides and oxyanions in members of the class Proteobacteria: characterization of tellurite, selenite, and rhodium sesquioxide reduction in *Rhodobacter sphaeroides*. *J. Bacteriol.*, Vol. 174, No. 5, (March 1992), pp. 1505-1514, ISSN 0021-9193

Moscoso, H., Saavedra, C., Loyola, C., Pichuantes, S. & Vásquez C. (1998). Biochemical characterization of tellurite-reducing activities of *Bacillus stearothermophilus* V. *Res. Microbiol.*, Vol. 149, No. 6, (June 1998), pp. 389-397, ISSN 0923-2508

O'Gara, J., Gomelsky, M. & Kaplan, S. (1997). Identification and molecular genetic analysis of multiple loci contributing to high-level tellurite resistance in *Rhodobacter sphaeroides* 2.4.1. *Appl. Environ. Microbiol.*, Vol. 63, No. 12, (December 1997), pp. 4713-4720, ISSN ISSN 0099-2240

Pérez, J.M., Calderón, I.L., Arenas, F.A., Fuentes, D.E., Pradenas, G.A., Fuentes, E.L., Sandoval, J.M., Castro, M.E., Elías, A.O. & Vásquez, C.C. (2007). Bacterial toxicity of potassium tellurite: unveiling an ancient enigma. *PLoS ONE*. Vol. 2, No. 2, (February 2007), pp. e211, ISSN 1932-6203

Saavedra, C.P., Encinas, M.V., Araya, M.A., Pérez, J.M., Tantaleán, J.C., Fuentes, D.E., Calderón, I.L., Pichuantes, S.E. & Vásquez, C.C. (2004). Biochemical characterization of a thermostable cysteine synthase from *Geobacillus stearothermophilus* V. *Biochimie*, Vol. 86, No. 7, (July 2004), pp. 481-485, ISSN 0300-9084

Silver S. (1996). Bacterial resistances to toxic metal ions--a review. *Gene*, Vol. 179, No. 1, (November 1996), pp. 9-19, ISSN 0378-1119

Silver, S. (2011). BioMetals: a historical and personal perspective. *Biometals*, Vol. 24, No. 3, (June 2011), pp. 379-390, OnlineISSN 1572-8773

Tai, C.H., Yoon, M.Y., Kim, S.K., Rege, V.D., Nalabolu, S.R., Kredich, N.M., Schnackerz, K.D. & Cook, P.F. (1998). Cysteine 42 is important for maintaining an integral active site for O-acetylserine sulfhydrylase resulting in the stabilization of the alpha-aminoacrylate intermediate. *Biochemistry*, Vol. 37, No. 30, (July 1998), pp. 10597-10604, ISSN 0006-2960

Tantaleán, J.C., Araya, M.A., Saavedra, C.P., Fuentes, D.E., Pérez, J.M., Calderón, I.L., Youderian, P. & Vásquez, C.C. (2003). The *Geobacillus stearothermophilus* V *iscS* gene, encoding cysteine desulfurase, confers resistance to potassium tellurite in *Escherichia coli* K-12. *J. Bacteriol.*, Vol. 185, No. 19, (October 2003), pp. 5831-5837, ISSN 0021-9193

Taylor, D.E. (1999). Bacterial tellurite resistance. *Trends Microbiol.*, Vol. 7, No. 3, (March 1999), pp. 111-115, ISSN 0966-842X

Tremaroli, V., Fedi, S. & Zannoni, D. (2007). Evidence for a tellurite-dependent generation of reactive oxygen species and absence of a tellurite-mediated adaptive response to oxidative stress in cells of *Pseudomonas pseudoalcaligenes* KF707. *Arch Microbiol.*, Vol. 187, No. 2, (February 2007), pp. 127-135, ISSN 1432-072X

Vásquez, C., Saavedra, C., Loyola, C., Araya, M. & Pichuantes, S. (2001). The product of the *cysK* gene of *Bacillus stearothermophilus* V mediates potassium tellurite resistance in *Escherichia coli*. *Curr. Microbiol.*, Vol. 43, No. 6, (December 2001), pp. 418-421, ISSN 1432-0991

Permissions

The contributors of this book come from diverse backgrounds, making this book a truly international effort. This book will bring forth new frontiers with its revolutionizing research information and detailed analysis of the nascent developments around the world.

We would like to thank David Figurski, for lending his expertise to make the book truly unique. He has played a crucial role in the development of this book. Without his invaluable contribution this book wouldn't have been possible. He has made vital efforts to compile up to date information on the varied aspects of this subject to make this book a valuable addition to the collection of many professionals and students.

This book was conceptualized with the vision of imparting up-to-date information and advanced data in this field. To ensure the same, a matchless editorial board was set up. Every individual on the board went through rigorous rounds of assessment to prove their worth. After which they invested a large part of their time researching and compiling the most relevant data for our readers. Conferences and sessions were held from time to time between the editorial board and the contributing authors to present the data in the most comprehensible form. The editorial team has worked tirelessly to provide valuable and valid information to help people across the globe.

Every chapter published in this book has been scrutinized by our experts. Their significance has been extensively debated. The topics covered herein carry significant findings which will fuel the growth of the discipline. They may even be implemented as practical applications or may be referred to as a beginning point for another development. Chapters in this book were first published by InTech; hereby published with permission under the Creative Commons Attribution License or equivalent.

The editorial board has been involved in producing this book since its inception. They have spent rigorous hours researching and exploring the diverse topics which have resulted in the successful publishing of this book. They have passed on their knowledge of decades through this book. To expedite this challenging task, the publisher supported the team at every step. A small team of assistant editors was also appointed to further simplify the editing procedure and attain best results for the readers.

Our editorial team has been hand-picked from every corner of the world. Their multi-ethnicity adds dynamic inputs to the discussions which result in innovative outcomes. These outcomes are then further discussed with the researchers and contributors who give their valuable feedback and opinion regarding the same. The feedback is then collaborated with the researches and they are edited in a comprehensive manner to aid the understanding of the subject.

Apart from the editorial board, the designing team has also invested a significant amount of their time in understanding the subject and creating the most relevant covers. They scrutinized every image to scout for the most suitable representation of the subject and create an appropriate cover for the book.

The publishing team has been involved in this book since its early stages. They were actively engaged in every process, be it collecting the data, connecting with the contributors or procuring relevant information. The team has been an ardent support to the editorial, designing and production team. Their endless efforts to recruit the best for this project, has resulted in the accomplishment of this book. They are a veteran in the field of academics and their pool of knowledge is as vast as their experience in printing. Their expertise and guidance has proved useful at every step. Their uncompromising quality standards have made this book an exceptional effort. Their encouragement from time to time has been an inspiration for everyone.

The publisher and the editorial board hope that this book will prove to be a valuable piece of knowledge for researchers, students, practitioners and scholars across the globe.

List of Contributors

J. Esclapez, M. Camacho, C. Pire and M.J. Bonete
Departamento de Agroquímica y Bioquímica, División de Bioquímica y Biología Molecular, Facultad de Ciencias, Universidad de Alicante, Alicante, Spain

David H. Figurski, Brenda A. Perez-Cheeks, Valerie W. Grosso, Karin E. Kram, Jianyuan Hua, Ke Xu and Jamila Hedhli
Department of Microbiology & Immunology, College of Physicians & Surgeons, Columbia University, New York, NY, USA

Daniel H. Fine
Department of Oral Biology, The University of Medicine & Dentistry of New Jersey, Newark, NJ, USA

Deepak Bastia, S. Zzaman and Bidyut K. Mohanty
Department of Biochemistry and Molecular Biology, Medical University of SC, Charleston, SC, USA

Luis Eduardo S. Netto
Instituto de Biociências – Universidade de Sao Paulo, Brazil

Marcos Antonio Oliveira
Universidade Estadual Paulista – Campus do Litoral Paulista, Brazil

Jürgen Ludwig, Holger Rabe, Anja Höffle-Maas, Marek Samochocki, Alfred Maelicke and Titus Kaletta
Galantos Pharma GmbH, Germany

Toni Petan and Jože Pungerčar
Department of Molecular and Biomedical Sciences, Jožef Stefan Institute, Ljubljana, Slovenia

Petra Prijatelj Žnidaršič
Department of Chemistry and Biochemistry, Faculty of Chemistry and Chemical Technology, University of Ljubljana, Ljubljana, Slovenia

Ewa Sajnaga, Ryszard Szyszka and Konrad Kubiński
Department of Molecular Biology, Institute of Biotechnology, The John Paul II Catholic University of Lublin, Poland

Jane E. Carland, Amelia R. Edington, Amanda J. Scopelliti, Renae M. Ryan and Robert J. Vandenberg
Department of Pharmacology, The University of Sydney, Australia

José Manuel Pérez-Donoso
Universidad de Chile, Chile

Claudio C. Vásquez
Universidad de Santiago de Chile, Chile